De Gruyter Graduate
Meyer, Reniers • Engineering Risk Management

Also of Interest

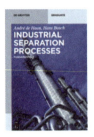

Industrial Separation Processes Fundamentals
De Haan, Bosch, 2013
ISBN 978-3-11-030669-9,
e-ISBN 978-3-11-030672-9

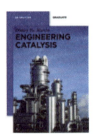

Engineering Catalysis
Murzin, 2013
ISBN 978-3-11-028336-5,
e-ISBN 978-3-11-028337-2

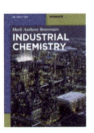

Industrial Chemistry
Benvenuto, 2014
ISBN 978-3-11-029589-4,
e-ISBN 978-3-11-029590-0

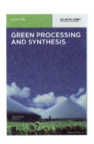

Green Processing and Synthesis
Hessel (Editor-in-Chief)
ISSN 2191-9542,
e-ISSN 2191-9550

Reviews in Chemical Engineering
Luss, Brauner (Editors-in-Chief)
ISSN 0167-8299,
e-ISSN 2191-0235

Thierry Meyer • Genserik Reniers

Engineering Risk Management

DE GRUYTER

Authors

MER Dr Thierry Meyer
Ecole Polytechnique
Fédérale de Lausanne
EPFL-SB-ISIC-GSCP
Station 6
1015 Lausanne
Switzerland
thierry.meyer@epfl.ch

Professor Genserik Reniers
University of Antwerp
Fac. of Applied Economic Science
Prinsstraat 13
02000 Antwerp
Belgium
genserik.reniers@ua.ac.be

This book has 122 figures and 40 tables

ISBN 978-3-11-028515-4
e-ISBN 978-3-11-028516-1

Library of Congress Cataloging-in-Publication Data
A CIP catalog record for this book has been applied for at the Library of Congress.

Bibliographic information published by the Deutsche Nationalbibliothek
The Deutsche Nationalbibliothek lists this publication in the Deutsche Nationalbibliografie; detailed bibliographic data are available in the Internet at http://dnb.dnb.de.

© 2013 Walter de Gruyter GmbH, Berlin/Boston.
The publisher, together with the authors and editors, has taken great pains to ensure that all information presented in this work (programs, applications, amounts, dosages, etc.) reflects the standard of knowledge at the time of publication. Despite careful manuscript preparation and proof correction, errors can nevertheless occur. Authors, editors and publisher disclaim all responsibility and for any errors or omissions or liability for the results obtained from use of the information, or parts thereof, contained in this work.

The citation of registered names, trade names, trade marks, etc. in this work does not imply, even in the absence of a specific statement, that such names are exempt from laws and regulations protecting trade marks etc. and therefore free for general use.

Typesetting: Compuscript Ltd., Shannon, Ireland
Printing and binding: Hubert & Co. GmbH & Co. KG, Göttingen

♾ Printed on acid-free paper
Printed in Germany

www.degruyter.com

About the authors

Thierry Meyer was born 1961 in Geneva. He obtained his MSc in Chemical Engineering at the Swiss Federal Institute of Technology in Lausanne (EPFL) in 1985, followed by a PhD in 1989, aslo in Chemical Engineering.

He joined Ciba-Geigy in 1994 as a development chemist in the pigment division, becoming head of development ad interim and went on to become the production manager in 1998.

In 1999 he switched to the Institute of Chemical Sciences and Engineering at EPFL, heading the Polymer Reaction Unit and then the chemical and physical safety research group.

Since 2005, he has also managed the occupational safety and health of the School of Basic Sciences at EPFL.

He serves as the Swiss academic representative in the working party on Loss Prevention and Safety Promotion of the European Federation of Chemical Engineering.

Genserik Reniers was born in 1974 in Brussels. He obtained an MSc in Chemical Engineering at the Vrije Universiteit Brussel in Brussels, and received his PhD in Applied Economic Sciences from the University of Antwerp, Belgium.

He founded the Antwerp Research Group on Safety and Security in 2006 at the University of Antwerp, coordinating multi- and inter-disciplinary safety and security research.

In 2007, he became responsible for safety and risk teaching and research at the Hogeschool – Universiteit Brussel, which forms part of the Association of the Catholic University of Leuven.

In 2012, he became responsible (in a part-time capacity) for teaching safety and risk management at the University of Antwerp.

He serves as the Belgian academic representative in the working party on Loss Prevention and Safety Promotion of the European Federation of Chemical Engineering.

Contents

1 Risk management is not only a matter of financial risk 1
References 6

2 Introduction to engineering and managing risks 7
2.1 Managing risks and uncertainties – an introduction 7
2.2 The complexity of risks and uncertainties 10
2.3 Hazards and risks 14
2.4 Simplified interpretation of (negative) risk 16
2.5 Hazard and risk mapping 19
2.6 Risk perception and risk attitude 21
2.7 ERM – main steps 23
2.8 Objectives and importance of ERM 29
2.9 Conclusions 30
References 30

3 Risk management principles 33
3.1 Introduction to risk management 33
3.2 Integrated risk management 35
3.3 Risk management models 37
    3.3.1 Model of the accident pyramid 37
    3.3.2 The P2T model 38
    3.3.3 The Swiss cheese model and the domino theory 39
3.4 The anatomy of an accident: SIFs and SILs 41
3.5 Individual risk, societal risk, physical description of risk 48
    3.5.1 Individual risk 48
    3.5.2 Societal risk 49
    3.5.3 Physical description of risk 52
        3.5.3.1 Static model of an accident 54
        3.5.3.2 Dynamic model of an accident 54
3.6 Safety culture and safety climate 56
    3.6.1 Organizational culture and climate 56
    3.6.2 Safety culture models 57
    3.6.3 The P2T model revisited and applied to safety and security culture and climate 60
3.7 Strategic management concerning risks and continuous improvement 62
3.8 The IDEAL S&S model 63
    3.8.1 Performance indicators 69
3.9 Continuous improvement of organizational culture 73
3.10 High reliability organizations and systemic risks 75

        3.10.1 Systems thinking.. 75
                3.10.1.1 Reaction time or retardant effect................................. 75
                3.10.1.2 Law of communicating vessels..................................... 75
                3.10.1.3 Non-linear causalities... 76
                3.10.1.4 Long-term vision.. 76
                3.10.1.5 Systems thinking conclusions....................................... 76
        3.10.2 Normal accident theory (NAT) and high reliability theory (HRT)... 76
        3.10.3 High reliability organization (HRO) principles............................ 79
                3.10.3.1 HRO principle 1: targeted at disturbances 79
                3.10.3.2 HRO principle 2: reluctant for simplification 79
                3.10.3.3 HRO principle 3: sensitive towards implementation 80
                3.10.3.4 HRO principle 4: devoted to resiliency 80
                3.10.3.5 HRO principle 5: respectful for expertise 80
        3.10.4 Risk and reliability.. 81
    3.11 Accident reporting... 82
    3.12 Conclusions... 83
    References... 84

4 Risk diagnostic and analysis 87
    4.1 Introduction to risk assessment techniques.. 87
        4.1.1 Inductive and deductive approaches... 88
        4.1.2 General methods for risk analysis 89
        4.1.3 General procedure 94
        4.1.4 General process for all analysis techniques................................. 96
    4.2 SWOT 98
    4.3 Preliminary hazard analysis 101
    4.4 Checklist... 103
        4.4.1 Methodology... 103
        4.4.2 Example 104
                4.4.2.1 Step 1a: Critical difference, effect of energies failures 104
                4.4.2.2 Step 1b: Critical difference, deviation from the
                            operating procedure. 105
                4.4.2.3 Step 2: Establish the risk catalogue.............................. 105
                4.4.2.4 Step 3: risk mitigation 106
        4.4.3 Conclusion.. 106
    4.5 HAZOP 107
        4.5.1 HAZOP inputs and outputs 108
        4.5.2 HAZOP process 108
        4.5.3 Example 111
        4.5.4 Conclusions 112
    4.6 FMECA 114
        4.6.1 FMECA inputs and outputs 115
        4.6.2 FMECA process 115
                4.6.2.1 Step 1: Elaboration of the hierarchical model,
                            functional analysis.. 116
                4.6.2.2 Step 2: Failure mode determination 117
                4.6.2.3 Step 3: The criticality determination........................... 119

	4.6.3 Example 119
	4.6.4 Conclusions 122
4.7	Fault tree analysis and event tree analysis 123
	4.7.1 Fault tree analysis 123
	4.7.2 Event tree analysis 126
	4.7.3 Cause-consequence-analysis (CCA): a combination of FTA and ETA 127
4.8	The risk matrix 130
4.9	Quantitative risk assessment (QRA) 135
4.10	Layer of protection analysis 138
4.11	Bayesian networks 140
4.12	Conclusion 144
References 145

5 Risk treatment/reduction 147
- 5.1 Introduction 147
- 5.2 Prevention 150
  - 5.2.1 Seveso Directive as prevention mean for chemical plants 151
  - 5.2.2 Seveso company tiers 155
- 5.3 Protection and mitigation 156
- 5.4 Risk treatment 159
- 5.5 Risk control 163
- 5.6 STOP principle 166
- 5.7 Conclusion 170
- References 170

6 Event analysis 173
- 6.1 Traditional analytical techniques 174
  - 6.1.1 Sequence of events 174
  - 6.1.2 Multilinear events sequencing 175
  - 6.1.3 Root cause analysis 175
- 6.2 Causal tree analysis 177
  - 6.2.1 Method description 177
  - 6.2.2 Collecting facts 178
  - 6.2.3 Building the tree 181
  - 6.2.4 Example 183
  - 6.2.5 Building an action plan 184
  - 6.2.6 Implementing solutions and follow-up 185
- 6.3 Conclusions 185
- References 186

7 Crisis management 187
- 7.1 Introduction 188
- 7.2 The steps of crisis management 190
  - 7.2.1 What to do when a disruption occurs 191
  - 7.2.2 Business continuity plan 193

	7.3	Crisis evolution.. 195
		7.3.1 The pre-crisis stage or creeping crisis.. 196
		7.3.2 The acute-crisis stage.. 196
		7.3.3 The post-crisis stage.. 196
		7.3.4 Illustrative example of a crisis evolution 197
	7.4	Proactive or reactive crisis management 198
	7.5	Crisis communication.. 199
	7.6	Conclusions.. 200
	References.. 200

8 Economic issues of safety 203
- 8.1 Accident costs and hypothetical benefits 204
  - 8.1.1 Quick calculation example of accident costs based on the number of serious accidents 206
- 8.2 Prevention costs... 208
- 8.3 Prevention benefits 208
- 8.4 The degree of safety and the minimum total cost point 209
- 8.5 Safety economics and the three different types of risks............................. 210
- 8.6 Cost-effectiveness analysis and cost-benefit analysis for occupational (type I) accidents 212
  - 8.6.1 Cost-effectiveness analysis.. 212
  - 8.6.2 Cost-benefit analysis.. 213
    - 8.6.2.1 Cost-benefit analysis for safety measures......................... 213
  - 8.6.3 Advantages and disadvantages of analyses based on costs and benefits.. 214
- 8.7 Optimal allocation strategy for the safety budget 215
- 8.8 Loss aversion and safety investments – safety as economic value 216
- 8.9 Conclusions.. 217
- References.. 217

9 Risk governance 219
- 9.1 Introduction... 219
- 9.2 Risk management system... 220
- 9.3 A framework for risk and uncertainty governance 226
- 9.4 The risk governance model (RGM).. 231
  - 9.4.1 The "considering?" layer of the risk governance model.............. 233
  - 9.4.2 The "results?" layer of the risk governance model 233
  - 9.4.3 The risk governance model.. 234
- 9.5 A risk governance PDCA... 234
- 9.6 Risk governance deficits 236
- 9.7 Conclusions.. 238
- References.. 239

10 Examples of practical implementation of risk management 241
- 10.1 The MICE concept 242
  - 10.1.1 The management step.. 243
  - 10.1.2 The information and education step 243

	10.1.3 The control step 244
	10.1.4 The emergency step 244
10.2	Application to chemistry research and chemical hazards 244
10.3	Application to physics research and physics hazards 246
	10.3.1 Hazards of liquid cryogens 247
	10.3.2 Asphyxiation 250
10.4	Application to emerging technologies 251
	10.4.1 Nanotechnologies as an illustrative example 254
10.5	Conclusions 256
	References 257

11 Major industrial accidents and learning from accidents 259
- 11.1 Link between major accidents and legislation 259
- 11.2 Major industrial accidents: Examples 261
  - 11.2.1 Feyzin, France, January 1966 261
  - 11.2.2. Flixborough, UK, June 1974 261
  - 11.2.3 Seveso, Italy, July 1976 262
  - 11.2.4 Los Alfaques, Spain, July 1978 263
  - 11.2.5 Mexico City, Mexico, November 1984 263
  - 11.2.6 Bhopal, India, December 1984 264
  - 11.2.7 Chernobyl, Ukraine, April 1986 264
  - 11.2.8 Piper Alpha, North Sea, July 1988 264
  - 11.2.9 Pasadena, Texas, USA, October 1989 265
  - 11.2.10 Enschede, The Netherlands, May 2000 265
  - 11.2.11 Toulouse, France, September 2001 266
  - 11.2.12 Ath, Belgium, July 2004 266
  - 11.2.13 Houston, Texas, USA, March 2005 266
  - 11.2.14 St Louis, Missouri, USA, June 2005 266
  - 11.2.15 Buncefield, UK, December 2005 267
  - 11.2.16 Port Wenworth, Georgia, USA, February 2008 267
  - 11.2.17 Deepwater Horizon, Gulf of Mexico, April 2010 268
  - 11.2.18 Fukushima, Japan, March 2011 268
- 11.3 Learning from accidents 268
- 11.4 Conclusions 271
- References 271

12 Concluding remarks 273

Index 277

1 Risk management is not only a matter of financial risk

Risk continues to perplex humankind. All societies worldwide, in the present and the past, have faced and face choices and decisions about how to adequately confront risks. Risk is indeed a key issue affecting everyone and everything. How to effectively and efficiently manage risks has been, is, and will always be, a central question for policy makers, industrialists, academics, and actually for everyone (depending on the specific risk). This is because the future cannot be predicted; it is uncertain, and no one has ever been successful in forecasting it. But we are very interested in the future, and especially in possible risky decisions and how they will turn out. We all face all kinds of risks in our everyday life and going about our day-to-day business. So would it not make sense to learn how to adequately and swiftly manage risks, so that the future becomes less obscure?

Answering this question with an engineering perspective is the heart of this book. If you went to work this morning, you took a risk. If you rode your bicycle, used public transportation, walked, or drove a car, you took a risk. If you put your money in a bank, or in stocks, or under a mattress, you took other types of risk. If you bought a lottery ticket you were involving an element of chance – something intimately connected with risk. If you decided to go ahead with one production process rather than another, you took a risk. If you decided to write a book, and not another book or some scientific papers, you took a risk. If you decided to publish one book, and not another one, you took a risk. All these examples reveal that "risk" may involve different meanings depending how we look at it.

The current highly competitive nature of economics might encourage firms to take on more or higher negative risks. But giving up managing such risks would be financially very destructive or even suicidal, even possibly in the short-term, because once disaster has struck it is too late to rewrite history. Some of you might argue that safety is expensive. We would answer that an accident is even much more expensive: the costs of a major accident are very likely to be huge in comparison with what should have been invested as prevention.

Let us take, as an example, the human, ecological and financial disaster of the Deepwater Horizon drilling rig in April 2010. Are the induced costs of several billions of euros comparable to the investments that could or might have averted the catastrophe? Answering this question is difficult, a priori, because it would require assessing these uncertainties and therefore managing a myriad of risks, even the most improbable. Most decisions related to Environment, Health and Safety (EHS) are based on the concept that there exists a low level of residual risk that can be deemed as "acceptably low". For this purpose, many companies have established their own risk acceptance or risk tolerance criteria. However, there exists many different types of risk and many methods of dealing with them, and at present, many organizations fail to do so. Taking the different types of risk into consideration, the answer to whether it would have been of benefit to the company to make all necessary risk management investments to prevent the Deepwater Horizon disaster, would have been – without any doubt – "yes".

In 2500 BC, the Chinese had already reduced risks associated with the boat transportation of grain by dividing and distributing their valuable load between six boats instead of one. The ancient Egyptians (1600 BC) had identified and recognized the risks involved by the fumes released during the fusion of gold and silver. Hippocrates (460–377 BC), father of modern medicine, had already established links between respiratory problems of stonemasons and their activity. Since then, the management of risks has continued to evolve:

Pliny the Younger (1st century AD) described illnesses among slaves.

In 1472, Dr. Ellenbog of Augsburg wrote an eight-page note on the hazards of silver, mercury and lead vapors (2).

Ailments of the lungs found in miners were described extensively in 1556 by Georg Bauer, writing under the name "Agricola" (3).

Dating from 1667 and resulting from the great fire that destroyed a part of London, the first Fire Insurance Act was published.

Although today's steep increase of risk- and risk management knowledge, and despite the neverending acceleration of all kinds of risk management processes, what still remains to be discovered in risk management and risk engineering is rather systemic and more complex.

As Ale (4) indicates, the essence of risk was formulated by Arnaud as early as 1662: "Fear of harm ought to be proportional not merely to the gravity of the harm, but also to the probability of the event". Hence, the essence of risk lies in the aspect of probability or uncertainty. Ale further notes that Arnaud treats probability more as a certainty than an uncertainty and as, in principle, measurable. Frank Knight even defines risk in 1921 as a "measurable uncertainty" (5). Today, the word "risk" is used in everyday speech to describe the probability of loss, either economic or otherwise, or the likelihood of accidents of some type. It has become a common word, and is used whether the risk in question is quantifiable or not. In Chapter 2 we will further elaborate on the true definition and the description of the concept "risk", and what constitutes it.

In response to threats to individuals, society, and the environment, policy makers, regulators, industry, and others involved in managing and controlling risks have taken a variety of approaches. Nowadays, the management of risk is a decision-making process aimed at achieving predetermined goals by reducing the number of losses of people, equipment and materials caused by accidents possibly happening while trying to achieve those goals. It is a proactive and reactive approach to accident and loss reduction. We will discuss risk management and its definition in a more general way in the next chapters, and we will define risk as having a positive and a negative side, but we will focus in the remainder of the book on managing risks with possibly negative consequences.

Human needs and wants for certainty can be divided into several classes. Abraham Maslow (6) discerned five fundamental types of human needs, and ordered them according to their importance in a hierarchical way in the form of a pyramid (▶Fig. 1.1).

As long as someone's basic needs at the bottom of the hierarchy are not satisfied, these needs demand attention and the other (higher) needs are more or less disregarded. "Safety", or in other words the strive for a decrease of uncertainty about negative risks, is a very important human need, right above basic needs such as food, drink, sleep and sex. Consequently, if risks are not well-managed in organizations, and people are not "safe",

Fig 1.1: Maslow's hierarchy of human needs.

the organizations will not be well-managed at all. They may focus upon "production" (the organizational equivalent of the physical basic needs), but the organizations will never reach a level in which they excel. Thus, engineering risk management (ERM), as discussed in this book, is essential for any organization's well-being and for its continuous improvement.

When reading this book it is necessary to wear "engineering glasses". This means that we will look at risk management using an engineer's approach – being systemic and not (necessarily) analytic.

> The analytic and the systemic approaches are more complementary than opposed, yet neither one is reducible to the other. In systemic thinking – the whole is primary and the parts are secondary; in analytic thinking – the parts are primary and the whole is secondary.
>
> The analytic approach seeks to reduce a system to its elementary elements in order to study in detail and understand the types of interaction that exist between them. By modifying one variable at a time, it tries to infer general laws that will enable us to predict the properties of a system under very different conditions. To make this prediction possible, the laws of the additivity of elementary properties must be invoked. This is the case in homogeneous systems, those composed of similar elements and having weak interactions among them. Here the laws of statistics readily apply, enabling us to understand the behavior of the multitude of disorganized complexity.
>
> The laws of the additivity of elementary properties do not apply in highly complex systems, like the risk management, composed of a large diversity of elements linked together by strong interactions. These systems must be approached by a systemic policy. The purpose of the systemic approach is to consider a system in its totality, its complexity, and its own dynamics (7).

This book is not written by financial experts, insurers, traders, bankers or psychologists. Although financial risk management is a crucial matter of daily business it will not be covered here. As a consequence, we will not discuss the topic of money investments, when it is more convenient to invest in bank accounts, shares or other financial placements. Our concern and purpose is to apply engineering methodologies to (non-financial) risk management.

What is the essential idea of risk management? We have all heard the saying, "Give a man a fish and you feed him for a day. Teach a man to fish, and you feed him for a lifetime". We could adapt this expression taking into account a risk management standpoint: "Put out a manager's fires, and you help him for a day. Teach a manager fire prevention, and you help him for a career". If a manager understands good risk management, he can worry about things other than fire-fighting.

Negative risk could be described as the probability and magnitude of a loss, disaster or other undesirable event. In other terms: something bad might happen. ERM can be described as the identification, assessment and prioritization of risks, followed by the coordinated and economical application of resources to minimize, monitor and control the probability (including the exposure) and/or impact of unfortunate events. In other terms: *being smart about taking chances.*

The first step is to identify the negative risks that a company faces: a risk management strategy moves forward to evaluate those risks. The simplest formula for evaluating specific risks is to multiply (after quantifying the risk in some form) the likelihood of the risky event by the damage of the event if it would occur. In other words, taking into consideration the possibility and consequences of an unwanted event.

The key word here is data. The best risk management specialists excel at *determining predictive data.*

The ultimate goal of risk management is to minimize risk in some area of the company relative to the opportunity being sought, given resource constraints. If the initial assessment of risk is not based on meaningful measures, the risk mitigation method, even if it could have worked, is bound to address the wrong problems. The key question to answer is: "How do we know it works?"

Several reasons could lead to the failure of risk management (8):

- The failure to measure and validate methods as a whole or in part.
- The use of factors that are known not to be effective (many risk management methods rely on human judgment and humans misperceive and systematically underestimate risks).
- The lack of use of factors that are known to be effective (some factors are proven to be effective both in a controlled laboratory setting and in the real world, but are not used in most risk management methods).

As already mentioned, risk is often measured by the likelihood (i.e., the probability in quantitative terms) of an event and its severity. Of the two, severity is more straightforward, especially after the event. Measuring the probability is where many people encounter difficulties. All we can do is use indirect measures of a probability, such as observing how frequently the event occurs under certain conditions or, when no information is available, make educated and expert assumptions.

Risk has to have a component of uncertainty as well as a cost (we have uncertainty when we are unable to quantify exactly accurate the likelihood).

Risk is generally thought of in highly partitioned subjects with very little awareness of the larger picture of risk management.

Risk is an uncertain event or set of circumstances which, should it or they occur, will have an effect on the achievement of objectives.

Very often we hear that experience helps in understanding risk, but:

Experience is a non-random, non-scientific sample of events throughout our lifetime.
Experience is memory based, and we are very selective regarding what we chose to remember.
What we conclude from our experience can be full of logical errors.
Unless we get reliable feedback on past decisions, there is no reason to believe our experience will tell us much.
No matter how much experience we accumulate, we seem to be very inconsistent in its application (8).

So at least there is need for a collective memory – the total memory of different experts with various expertise backgrounds.

There is space for improving risk management so as to adopt the attitude of modeling uncertain systems: acting as a scientist in choosing and validating models; building the community as well as the organization. Risk management is not a sole matter of managers or risk experts, because it involves and affects everyone in the organization, and directly or indirectly a lot of people who form no part of the organization.

Near-misses (being interpreted as something that "only just did not happen", but was very close to happening) tend to be much more plentiful than actual disasters and, as at least some of them are caused by the same event that caused the near-miss, we can learn something very useful from them.

The evolution of risk management has been influenced by expanding knowledge and tools as well as by the hazards that need to be addressed. Regulatory bodies, which tend to react in response to incidents, over time enacted measures to prevent recurrences. These bodies also have shaped how hazards are identified and controlled.

This book intends to provide the reader with the necessary background information, ideas, concepts, models, tools, and methods that should be used in a general organized process to manage risks. The reader will discover that risk management is considered to be a process which affords assurance that:

Objectives are more likely to be achieved.
Unwanted events will not happen or are less likely to happen.
Goals will be – or are more likely to be – achieved.

The engineering risk management process enables the identification and evaluation of risks, helps in setting acceptable risk thresholds, ranks them, allows for the identification and mapping of controls against those risks, and helps identify risk indicators that give early warning that a risk is becoming more serious or is crystallising. Once the risk management process has been followed and controlled, a risk recording has been produced and risk owners have been identified, the risk should be monitored and reviewed. Risk management is a *never-ending process*, being more iterative than ever with the increasingly fast evolution of emerging technologies.

Every organization has to live with negative risks. They go hand-in-hand with the positive risks that lead to gains. Managing those negative risks in an adequate manner is therefore a vital element of good governance and management. Several types and sources of risks – either internal or external to the activity – may affect a business project, process, activity, etc. Some risks may be truly unpredictable and linked to large-scale structural causes beyond a specific activity. Others may have existed for a long time or may be foreseen to occur in the future.

As already indicated, the key idea of risk is that there is uncertainty involved. If compared with life (9), the only certainty in life is death, and the uncertainty lies in when and how death occurs. People strive to delay the final outcome of life and try to improve the quality of life in the interim. Threats to these interim objectives involve risks, some natural, some man-made, some completely beyond our control, but most of them controllable and manageable.

In summary, it is not always obvious whether something is good/right or bad/wrong, as is often the case in movies. Some action can be bad for one person, but it can be good for another. Or something can be wrong at first sight, but it can be right in the long run. Or it can be bad for few, but good for many, etc. Deciding whether some action or process is good or bad is thus more complex than it seems on the surface. Hence, life is not black or white, as it is so often pictured in stories and fairy tales. Life is colorful and complex. Dealing with life is complex. Risks are not only a natural part of life, they can even be compared with life. Dealing with risks is much more complex than appears at first view: there is a positive and negative side to most risks and thus they are not black or white. Handling risks is therefore not easy or straightforward. Uncertainties in life are the very same as those related to risks. Decisions in life or in organizations are comparable with those related to risks. Making the right decisions in life (which is often very difficult because of the complex character of the decision-making process, missing information, etc.) leads to a prosperous and possibly longer life. Making the right decisions in business (which is also difficult because of the complexity of the problems at hand and the uncertainties accompanying available information) leads to sustainable profits and healthy organizations.

To make the right decisions time and time again, risks should be "engineered"; that is, they should be managed with engineering principles. To help anyone make the right decisions in life and in business, and thereby continuously improve his/her/its position, is what this book is written for.

References

1. Ramazzini, B. (1713) De morbis artificum deatriba, trad. Diseases of Workers. Translated by Wright W.C., Chicago, IL: University of Chicago Press, 1940.
2. Rosen, G. (1976) A History of Public Health. New York: MD Publications.
3. Agricola, G. (1556) De Re Metallica T. 12th edn, 1912, Translated by Hoover H.C. and Hoover L.H. London: The Mining Magazine.
4. Ale, B.J.M. (2009) Risk: An Introduction. The Concepts of Risk, Danger and Chance. Abingdon, UK: Routledge.
5. Knight, F.H. (1921) Risk, Uncertainty and Profit. 2005 edn. New York: Cosimo Classics.
6. Maslow, A.H. (1943) A theory of human motivation. Psychol. Rev. 50:370–396.
7. Umpleby, S.A., Dent, E.B. (1999) The origins and purposes of several traditions in systems theory and cybernetics. Cybernet Syst. 30:79–104.
8. Hubbard, D.W. (2009) The Failure of Risk Management. Hoboken, NJ: John Wiley & Sons Inc.
9. Rowe, W.D. (1977) An Anatomy of Risk. New York: John Wiley & Sons Inc.

2 Introduction to engineering and managing risks

2.1 Managing risks and uncertainties – an introduction

Traditionally, two broad categories of management systems can be distinguished: business management systems and risk management systems. The former are concerned with developing, deploying and executing business strategies, while the latter focus on reducing safety, health, environmental, security, and ethical risks.

Business management systems specifically aim at improving the quality or business performance of an organization, through the optimization of stakeholder satisfaction, with a focus on clients – such as the ISO Standard 9001:2008 (1) – or extended to other stakeholders (e.g., employees, society, shareholders, etc.) – such as the EFQM 2010 Model for Business Excellence (2) or the ISO 9004:2009 Guidelines (3).

Some of the most popular generic examples of risk management systems are the international standard for environmental management [ISO 14001:2004 (4)], the European Eco-Management and Audit Scheme EMAS (5), the internationally acknowledged specification for occupational safety and health [OHSAS 18001:2007 (6)], and the international standard for integrity management SA 8000 (7).

The boundaries of those two categories have seemed to fade in recent years. Firstly, through the increase of alignment and compatibility between the specific management systems, as a result of successive revisions (8). Secondly, by the introduction of a modular approach with a general, business-targeted, framework, and several sub-frameworks, dealing with specific issues, such as risk management, as is shown by the development of the EFQM Risk management model in 2005. Thirdly, this evolution is illustrated by the emergence of integrated risk management models, which emphasize both sides of risks: the threat of danger, loss or failure (the typical focus of risk management) as well as the opportunity for increased business performance or success (the typical focus of business management). Examples of this last category are the Canadian Integrated Risk Management Framework (2001) (9) and the Australian-New Zealand standard AS/NZS 4360:2004 (10), which served as the basis for the development of the generic ISO Risk management standard 31000:2009 (11). It is interesting to consider the latest insights of ISO 31000:2009 to discuss the risk concept.

A *risk* is defined by ISO 31000:2009 as "the effect of uncertainties on achieving objectives" (ISO, 2009) (11). Our world can indeed not be perfectly predicted, and life and businesses are always and permanently exposed to uncertainties, which have an influence on whether objectives will be reached or not. The only certainty there is about risks, is that they are characterized with uncertainty. All other features result from assumptions and interpretations.

The ISO 31000:2009 definition implies that risks (financial as well as non-financial, technological) are two-sided: we call them negative risks if the outcome is negative and positive risks if the outcome is positive, although the same risks are envisioned (▶Fig. 2.1). It is a straightforward assumption that organizations should manage risks in

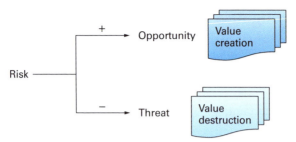

Fig. 2.1: Negative or positive risk.

a way that the negative outcomes are minimized, and that the positive outcomes are maximized. This is called risk management (RM) and contains, amongst others, a process of risk identification, analysis, evaluation, prioritization, handling, and monitoring (see below) aimed at being in control of all existing risks, whether they are known or not, and whether they are positive or negative.

Continuous risk management lowers the risk of disruption and assesses the potential impacts of disruptions when they occur. It has been developed over time and within many sectors in order to meet diverse needs. The adoption of consistent processes within a comprehensive framework can help to ensure that all types and numbers of risk are managed effectively, efficiently and coherently across an organization. In current industrial practice, risk management is only focused on negative risks, and only on avoiding losses, instead of simultaneously avoiding losses and producing gains. This is mainly because companies tend to consider "risk managers" as "negative risk managers" for historical reasons: risk managers have been appointed in organizations mainly to satisfy legislative requirements or because of incidents and accidents that happened within firms; hence the only risks that needed to be managed displayed possible negative consequences. Nonetheless, to take all aspects of risks into account and to take optimal decisions, risks should thus ideally be viewed from a holistic viewpoint, meaning that all relevant stakeholders and experts should be involved in the risk management process, and that all possible knowledge and know-how should be present.

The best available "classic viewpoint" of risk management is one of (narrow) thinking of purely organizational risk management, and taking different domains within the company into account: integrating them to take decisions for maximizing the positive side of risks and minimizing the negative side. The process for taking decisions in organizations should however be much more holistic than this. Risks are characterized by internal as well as external uncertainties. Hence, all these uncertainties and their possible outcome(s) should be anticipated, that is, identified and mapped, for every risk, and by different types of experts and stakeholders. If this can be achieved, the optimal decisions can be taken. The end goal is to use all the right people and means, at the right time, to manage all existing risks in the best possible way, whether the risks are positive or negative, or whether they are known or not. After all, a part of risk management is improving our perception of reality and "thinking about the unthinkable".

To manage uncertainties efficiently, a composite of three building blocks (knowledge and know-how, stakeholders and expertise, and mindset) can be conceptualized, as illustrated in ▶Fig. 2.2.

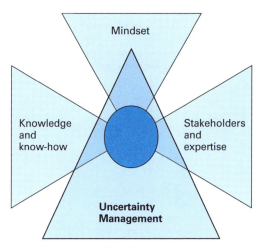

Fig. 2.2: Required capabilities for adequate uncertainty management.

For each of the building blocks needed by an organization to efficiently perform "uncertainty management", some recommendations can be suggested. To deal with the knowledge and know-how needs, cognition should be present in the organization about past, present and (e.g., scenario-based) future data and information, and about risk taking, risk averting, and risk neutral items. Furthermore, the organization should collect information on laws and regulations, rules and guidelines, and best practices and ideas; and also on the procedural, technological and people domains. To address the stakeholders and expertise building block, involvement should be considered from different organizations, authorities and academia, as well as from other stakeholders (clients, personnel, pressure groups, media, surrounding communities, etc.), and from different types of disciplines (engineers, medical people, sociologists, risk experts, psychologists, etc.), and used where deemed interesting. The mindset building block indicates that, besides existing principles, some additional principles should be followed for adequate uncertainty management: circular, non-linear, and long-term thinking, law of iterated expectations, scenario building, and the precautionary principle, and operational, tactic, and strategic thinking. All these requirements should be approached with an open mind.

Risk management can actually be compared with mathematics: both disciplines are commonly regarded as "auxiliary science" domains, helping other "true sciences" to get everything right. In the case of mathematics, true sciences indicate physics, chemistry, bio-chemistry, and the like: without mathematics, these domains would not be able to make exact predictions, conclusions, recommendations, etc. The same holds for risk/uncertainty management: without this discipline, applied physics, industrial chemistry, applied bio-chemistry, etc. lead to sub-optimal results and achieving objectives will be hampered. Hence, mathematics is needed for correct laws in physics, chemistry, etc., and risk management is required for optimized applications in physics, chemistry, etc. ▶Fig. 2.3 illustrates this line of thought.

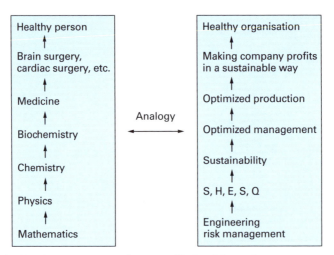

Fig. 2.3: Analogy between exact sciences and industrial practice.

Bearing ▶Fig. 2.3 in mind, although the management of negative risks has proven to be extremely important over the past decades to save lives and/or to avoid illnesses and injuries in all kinds of organizations, major and minor accidents as well as occupational illnesses are still very much present in current industrial practice. Hence, there is still room for largely improving all management aspects of negative risks and it can be made much more effective. How this can be theoretically achieved, is treated in this book; we thus mainly focus on theories, concepts, techniques, models, frameworks, and practical examples related to unwanted events leading to undesirable consequences, in other words, on the field of management of negative uncertainties and risks.

> Uncertainties and risks cannot be excluded out of our lives. Therefore, we should always be able to deal with them in the best possible way. To this end, a risk should be viewed as a two-sided coin: one side is positive, and the other side is negative. Baring this in mind, we need the right mind-set, enough information, and the right people to deal with risks and uncertainties.

2.2 The complexity of risks and uncertainties

"Risk" means different things to different people at different times. However, as already mentioned, one element characterizing risk is the notion of uncertainty. That is, for the omniscient or omnipotent the "risk" concept would be incomprehensible. Nonetheless, in the real world, the future of events, handlings, circumstances, etc. cannot be perfectly predicted. Unexpected things happen and cause unexpected events. An "uncertainty sandglass" can thus be drafted; and strengths, weaknesses, opportunities, and threats (SWOT, see also Chapter 4, Section 4.2) can be identified. ▶Fig. 2.4 displays the uncertainty sandglass, with the SWOT elements situated within the concept.

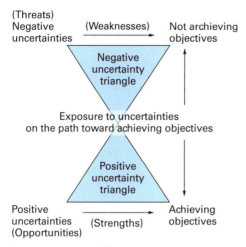

Fig. 2.4: The uncertainty sandglass.

A SWOT analysis is a strategic planning method that specifies the objectives of, e.g., a business venture or a project and aims to identify the internal and external factors that are favorable and unfavorable to achieve those objectives. Using the principles of a SWOT analysis (identifying positive as well as negative uncertainties) allows risk managers to close the gap between line management and corporate or top management regarding taking decisions on existing uncertainties in the organization. By not only focusing on the negative side of risks (weaknesses and threats), but by also emphasizing the positive side of risks (strengths and opportunities), an approach is adopted by the risk manager that is more easily understandable by top management.

As mentioned above, in this book we only focus on the top triangle of the uncertainty sandglass: negative uncertainties – exposure to uncertainties – not achieving objectives. If we translate this into "(negative) risk management language", the upper triangle should be: hazards – exposure to hazards – losses. This triangle can be called the "Risk Trias". If one of these elements are removed from this triangle ("trias" in Latin), there is no risk. The engineering aspects of risk management, discussed in this book, focus on how to diminish, decrease or soften as much as possible one of the three elements of the "Risk Trias" (hazards, exposure, or losses), or a combination thereof.

Roughly, three types of uncertainties can be identified:

- Type I – uncertainties where a lot of historical data is available.
- Type II – uncertainties where little or extremely little historical data is available.
- Type III – uncertainties where no historical data is available.

From the viewpoint of negative risks, consequences of type I uncertainties mainly relate to individual employees (e.g., most work-related accidents). The outcome of type II uncertainties may affect a company or large parts thereof, e.g., large explosions, internal domino effects, etc. – for this sort of accident the reader is referred to Lees (12), Wells (13), Kletz (14,15), Atherton and Gil (16) and Reniers (17). Type III uncertainties, e.g., have an unprecedented and unseen impact upon the organization and society.

Thus, whereas type I negative risks lead to most work-related accidents, such as falling, little fires, slipping, etc., type II negative risks can result in catastrophes with major consequences and often with multiple fatalities. Type II accidents do occur on a (semi-) regular basis in a worldwide perspective, and large fires, large releases, explosions, toxic clouds, etc. belong to this class of accident. Type III negative risks may transpire into "true disasters" in terms of the loss of lives and/or economic devastation. These accidents often become part of the collective memory of humankind. Examples include disasters such as Seveso (Italy, 1976), Bhopal (India, 1984), Chernobyl (USSR, 1986), Piper Alpha (North Sea, 1988), 9/11 terrorist attacks (USA, 2001), and more recently Deepwater Horizon (Gulf of Mexico, 2010) and Fukuchima (Japan, 2011).

For preventing type I risks turning into accidents, risk management techniques and practices are widely available. We will discuss some of those techniques in Chapter 4. Statistical and mathematical models based on past accidents can be used to predict possible future type I accidents, indicating the prevention measures that need to be taken.

Type II uncertainties and related accidents are much more difficult to predict via commonly used mathematical models because the frequency with which these events happen is too low and the available information is not enough to be investigated via e.g., regular statistics. The errors of probability estimates are very large and one should thus be extremely careful while using such probabilities. Hence, managing such risks is based on the scarce data that are available and on extrapolation, assumption and expert opinion, e.g., see Casal (18). Such risks are also investigated via available risk management techniques and practices, but these techniques should be used with much more caution as the uncertainties are much higher for these than for type I risks. A lot of risks (and latent causes) are present that never turn into large-scale accidents because of adequate risk management, but very few risks are present that turn into accidents with huge consequences. Hence, highly specific mathematical models (such as QRAs, see Chapter 4, Section 4.9) should be employed for determining such risks. It should be noted that the complex calculations lead these risks to appear accurate; nonetheless, they are not accurate at all (because of all the assumptions that have to be made for the calculations) and they should be regarded and treated as relative risks (instead of absolute risks). As Balmert (19) indicates, physicist Richard Feynman's dissenting opinion in the Rogers Commission Report (20) offers the perfect perspective: by NASA's risk management protocols, perhaps among the most sophisticated work on risk, the failure rate for the Space Shuttle orbiter was determined to be one in 125,000. Feynman used a common-sense approach, asking scientists and engineers the approximate failure rate for an unmanned rocket. He came up with a failure rate of between 2% and 4%. With two failures – *Challenger* and *Columbia* – in 137 missions, Feynman's "back-of-the-envelope" calculation was by far the more accurate.

Type III uncertainties are extremely high, and their related accidents are simply impossible to predict. No information is available about them and they only happen extremely rarely. They are the result of pure coincidence and they cannot be predicted by past events in any way, they can only be predicted or conceived by imagination. Such accidents can also be called "black swan accidents" (21). Such events can truly only be described by "the unthinkable" – which does not mean that they cannot be thought of, but merely that people are not capable of (or mentally ready for) realizing that such events really may take place. ▶Fig. 2.5 illustrates in a qualitative way the three uncertainty-types of events as a function of their frequency.

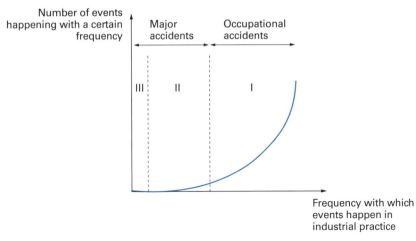

Fig. 2.5: Number of events as a function of the events' frequencies (qualitative figure).

Type I unwanted events from ▶Fig. 2.5 can be regarded as "occupational accidents" (e.g., accidents resulting in the inability to work for several days, accidents requiring first aid, etc.). Type II and type III accidents can both be categorized as "major accidents" (e.g., multiple fatality accidents, accidents with huge economic losses, etc.).

Another way of explaining this distinction between possible events is by using a slightly adapted table from the statement of Donald Rumsfeld on the absence of evidence linking the government of Iraq with the supply of weapons of mass destruction in terrorist groups (US DoD, 2002) (22). ▶Fig. 2.6 illustrates the different types of events.

Events are classified into different groups, based on the one hand on the available knowledge or information on the events from the past, and on the other hand on the fact that people have an open mind towards the possibility of the event. Looking at the

	Knowledge	Lack of knowledge
Open mind	Known known	Known unknown
Closed mind	Unknown known	Unknown unknown

Fig. 2.6: Table differentiating between known and unknown.

knowledge of the events from the past, or the lack of it, in combination with the level of open mindedness of people towards the event, four separate groups can be distinct:

1. Events that we do not know from the past (they have never occurred), and look at with a closed mind. These events are called "Unknown unknowns".
2. Events that we know from the past (we have certain information or records about them), and look at with a closed mind. These events are called "Unknown knowns".
3. Events that we know from the past and have an open mind towards. These events are called "Known knowns".
4. Events that we do not know from the past, but which we have an open mind towards. They are defined as "Known unknowns".

It should be clear that type III events are classified as Unknown unknowns, e.g., Fukushima (2011) and Deepwater Horizon (2010). Type II events can be regarded as Unknown knowns, e.g., Buncefield (2005) and Toulouse (2001) and Known unknowns, e.g., emerging technologies such as nanotechnology. Type I events are obviously the Known knowns. (See Chapter 11 for discussion of these and other accidents.) Hence, an open mind and sufficient information are equally important to prevent all types of accidents from occurring.

> Three types of negative risks exist: type I encompasses all risks where a lot of information is available, type II encompasses all risks where only very scarce information is available, and type III encompasses all risks where no information is available.

2.3 Hazards and risks

From the "Risk Trias", it is obvious that an easy distinction can be made between what is a *hazard* and what is a *risk*.

A *hazard* can be defined as "The potential of a human, machine, equipment, process, material, or physical factor to lead to an unwanted event possibly causing harm to people, environment, assets or production". Hence, a hazard is a disposition, condition or situation that may be a source of an unwanted event leading to potential loss. Disposition refers to properties that are intrinsic to a hazard and will be harmful under certain circumstances.

Although *risk* is a familiar concept in many fields and activities including engineering, law, economics, business, sports, industry, and also in everyday life, various definitions exist. To link with the "Risk Trias", the following definition of *(negative) risk* can be coined: "The possibility of loss (injury, damage, detriment, etc.) created by exposure to one or more hazards. The significance of risk is a function of the likelihood of an unwanted event and the severity of its consequences." The event likelihood may be either a frequency (the number of specified events occurring per unit of time, space, etc.), a probability (the probability of a specified incident following a prior incident), or a qualitative expression, depending on the circumstances and the information. The magnitude of the loss determines how the risk is described on a continuous scale from diminutive to catastrophic.

Kirchsteiger (23) points out that in reality risk is not simply a product type of function between likelihood and consequence values, but an extremely complex multi-parametric function of *all* circumstantial factors surrounding the event's source of occurrence. However, in order to be able to make a reproducible and thorough risk assessment, and since a risk can be regarded as a hazard that has been quantified, it is important to establish a formula for the quantification of any hazard. In the currently accepted definition, risk is calculated by multiplying the likelihood of an unwanted event by the magnitude of its probable consequences:

> Negative risk = (Likelihood of unwanted event) × (Severity of consequences of unwanted event)

Thus, if we are able to accurately and quantitatively assess the likelihood of an event, as well as the probable severity of the event's consequences, we find a quantitative expression of risk. A risk can thus be expressed by a "number". It should be noted that the *exposure* factor of the Risk Trias is part of the likelihood assessment. As Casal (18) points out, such an equation of risk is very convenient for many purposes, but it also creates several difficulties, e.g., determining the units in which risk is measured. Risk can be expressed in terms of a number of fatalities, the monetary losses per unit of time, the probability of certain injuries to people, the probability of a certain level of damage to the environment, etc. Also, in some cases, especially in type II and III events, it is obviously very difficult to estimate the likelihood of a given unwanted event and the magnitude of its consequences. To be able to obtain adequate risk assessments, appropriate methods and approaches are used, as explained further in this book.

Examples of the different Risk Trias elements are given in ▶Tab. 2.1.

▶Tab. 2.1 clearly demonstrates that if either the hazard, the exposure, or the loss are avoided or decreased, the risk will diminish or even cease to exist. Hence, risk management techniques are always aimed at trying to avoid or decrease one or several of the Risk Trias elements. The usefulness of risk analysis and risk management is also easy to understand. An adult working with a fryer will be less risky than a child working with it. Prevention measures are thus obviously more important in case of the latter situation. Also, a trained and competent worker using the toxic material will lead to a lower risk than an unskilled and incompetent worker. In the same way, an automated system will lead to lower risk levels than a human-operated system, in the case of look-alike product storage.

Tab. 2.1: Illustrative examples of different concepts of Risk Trias.

Example	Hazard	Exposure	Loss	Risk
Fryer in use	Heat	Person working with the fryer	Burns	Probability of children having burns with a certain degree
Toxic product	Toxicity	Person working with the toxic product	Intoxication	Probability of worker being intoxicated with a certain severity
Storage of products that look alike	Looking alike	Order by client of one of the look-alike products	Wrong delivery to client	€10,000 claim by client over a certain period of time

Furthermore, it is essential to realize that *a risk value is always relative*, never absolute, and that risk values should always be compared with other risk values, to make risk management decisions. Risks should thus be viewed relative to each other – always.

Moreover, by looking at the likelihood values and the consequence values, and taking all the many aspects into consideration, it is possible to put different kinds of risk in proper perspective. This is important as the risk value may depend on the viewpoint from which it is looked at. For example, there often remains a gap between the technical and the non-technical view of risk (24). As an example, if the safety record of an individual airplane is, on average, one accident in every 10,000,000 flights, engineers may deem air travel to be a reliable form of transport with an acceptable level of risk. However, when all the flights of all the airplanes around the world are accumulated, the probability is that several airplane crashes will occur each year – a situation borne out of observed events. Each crash, however, is clearly unacceptable to the public, in general, and to the people involved in the crashes, in particular.

2.4 Simplified interpretation of (negative) risk

From the above, it is clear that risk is a very theoretical concept and cannot be felt by human senses; it can be only measured or estimated. Risk possesses a value – its criticality – which can be estimated by the association of the constitutive elements of risk in a mathematical model. Always considering that, we stay in the area of its probability.

The simplest model defines that the probability of a certain risk depends:

- on the frequency by which the target is exposed to the hazard (sometimes called likelihood of occurrence), supposed that the hazard threatens the target, and
- the evaluation of its consequence corresponding to a measurement of the severity of the mentioned consequences.

This model is generally expressed by:

$$\text{Risk} = \text{frequency} \times \text{severity}; \quad R = F \cdot G$$

To be consistent with the concept that measures have been taken against a threat, the notions of protection and prevention are taken into account in the following formula:

$$R = F \cdot G = \left(\frac{N \cdot T}{Pre}\right) \cdot \left(\frac{D}{Pro}\right) \qquad \text{Eqn. (2.1)}$$

The likelihood of occurrence F depends on:

- N – number of set targets
- T – average exposure time of each target at risk
- Pre – prevention implemented to reduce N or T

Severity G is function of:

- D – "crude" hazard of the situation
- Pro – level protection implemented in the light of this hazard

2.4 Simplified interpretation of (negative) risk | 17

This formula indicates the possible pathways and solutions to reduce risk. Reducing risk means to act on:

- The severity → reduce the hazard, increase protection measures.
- The frequency → reduce the exposure time, reduce the number of exposed targets, increase prevention measures.

Changing the risk by reducing the occurrence of its components and/or their severity means addressing the following questions:

- Is it possible to reduce the number of exposed targets (*N*)?
- Is it possible to reduce the time the targets are exposed to the hazard (*T*)?
- Is it possible to increase the prevention measurements (*Pre*)?
- Is it possible to reduce the hazardousness (*D*)?
- Is it possible to increase the level of protection (*Pro*)?

Zero risk does not exist, except if the hazard is zero or non-existent or if there is no exposition to the hazard.

The results of the evaluation of risk are often presented in a matrix-like manner. An illustrative example is depicted in ▶Fig. 2.7, the terms of likelihood of occurrence and the severity of the damages are easy to understand. The analyzed situation is quantified in terms of frequency and severity and placed in the corresponding cell (see also Chapter 4, Section 4.8).

Thereby the quantification of risk does not solely indicate something about the occurrence and severity. It is not easy to make a risk very concrete because it is not comprehensible. However, with the following illustration it is possible to understand the different classes of risk. Let us consider the drop of a spherical object – there are several possibilities (▶Fig. 2.8):

- A – low probability, low damage = low risk
- B – high probability, low damage = medium risk
- C – low probability, high damage = medium risk
- D – high probability, high damage = high risk

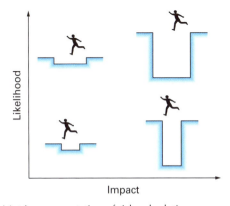

Fig. 2.7: Matrix representation of risk calculation.

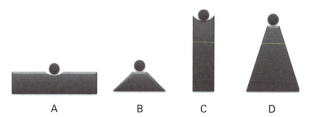

Fig. 2.8: Model of risk classes for a falling spherical object.

In order to be even more concrete, let us illustrate the risk by a concrete example: a fall from a cliff (▶Fig. 2.9). We could note that the parameters present in Eq. (2.1) can be materialized as (only a few examples are presented):

- D = height of the cliff, ground, slippery ground on top of the cliff
- Pre = level of protective fences, warnings signs
- Pro = being secured with a rope, having a loose soil, e.g., water
- T = average time for a walk along the cliff
- N = number of people walking along the cliff

The risk of a person falling will depend on:

- If the surface is slippery or not
- If there are fences or guardrails
- If the person is secured or not
- Aggravating factors, etc.

A cliff might be very risky and:

(i) a little dangerous if:
 - the grass surface is slippery
 - there are no guardrails
 - the cliff height is low
 - there is loose soil at the cliff bottom.
(ii) very dangerous when:
 - the cliff is without fences, or
 - the cliff is high (even if there are warning signs), or
 - there is hard ground at the bottom (rocks instead of sand or water).

We could also imagine that some worsening factors can play a role, e.g., if mist is present and obscuring the edge of the cliff; a strong wind could also worsen that risk. This example emphasizes that even if the reduced formula allows, estimating the risk may be simple, its interpretation and real estimation may be difficult.

> To exist, risks need three factors: hazards, losses, and exposure. Decreasing (or even avoiding or taking away) one of these parameters, or (one of) the characteristics of these parameters, in one way or another, leads to lower risks (or even to no risk).

2.5 Hazard and risk mapping | 19

Fig. 2.9: Image of a cliff, to illustrate risk determination.

2.5 Hazard and risk mapping

Many risk analysis techniques and risk management approaches emerged in the industry from the 1960s onwards. This can be regarded as a reaction to some major accidents as well as the desire to achieve higher performance, and to improve production, quality, and workers' health.

A hazard map highlights areas that are affected by or vulnerable to a particular hazard. The objective of hazard maps is to provide people with the information on the range of possible damage and the accident prevention activities. It is important that people are provided with understandable clear information. For example we could illustrate this by natural disaster prevention. This is necessary to protect human lives, properties and social infrastructure against disaster phenomena. Two types are possible:

- Resident-educating: this type of map aims to inform the residents living within the damage forecast area of the risk of danger. The information on areas of danger or places of safety and the basic knowledge on disaster prevention are given to residents. Therefore, it is important that such information is represented in an understandable form.
- Administrative information: This type of map is used as the basic materials that administrative agencies utilize to provide disaster prevention services. These hazard maps can be used to establish a warning system and the evacuation system, as well as evidence for land-use regulations. They may also be used in preventive works.

Hazard mapping provides input to educational programs to illustrate local hazards, to scientists studying hazard phenomena, land-use planners seeking to base settlement locations to reduce hazard impacts, and to combine with other information to illustrate community risks. A map of each type of hazard has unique features that need to be displayed to provide a picture of the distribution of the variations in the size and potential severity of the hazard being mapped. While field work provides data input for many hazard maps, some data will need to be collected from past impacts, from archives maintained of records collected from beneath the built environment, from instruments, and/or from the analysis of other data to develop a model of future impacts. The methods for collecting data (e.g., surveys, etc.) are still important in providing input to producing hazard maps. Understanding potential hazards typically involves identifying adverse conditions (e.g., steep slopes) and hazard triggers (e.g., rain fall). Together, conditions and triggers help define the potential hazard.

The main difficulty in mapping hazards is to quantify the intensity of the hazard. Going back to hazards found in the workplace, Marendaz et al. (25) established a comprehensive list of specific hazards found in research institutions. The strength of this platform (ACHiL) is that it gives clear and objective criteria for each hazard (comprising 27 categories). This allows for the reduction of the effects of incorrect hazard perception by individuals. The ACHiL platform identifies and classifies hazards according to a four-level scale:

- 0 – Hazard is not present.
- 1 – Moderate hazard.
- 2 – Medium hazard.
- 3 – Severe hazard.

Depending on each hazard, quantitative thresholds have been settled from level 1 to 3. This also includes the applied criteria used for the levels discrimination. Hence a non-qualified person in health and safety would be less likely to make mistakes when assessing hazards. An illustration of the hazard mapping for a research building is presented in ▶Fig. 2.10. It is obvious that very easily we can observe where the most hazardous zones are located and which are the hazards concerned.

Another scope of the hazard mapping or hazard portfolio is to identify where resources for an in-depth risk analysis should be performed. Hence, as resources are not infinite, it would make sense to concentrate on where hazards are most important, or where different hazards could, in a synergetic way, lead to combined hazards being often less

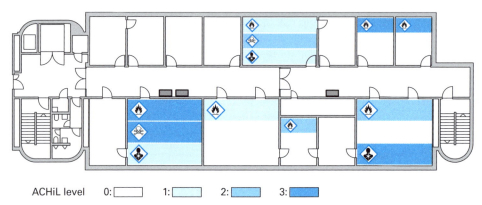

Fig. 2.10: Hazard mapping of a lab sector for flammable solvents, acute and chronic toxics after ACHiL scale application.

obvious to note. In addition, hazard mapping is a powerful decision support tool, which can be used to detect rapidly "hot spots", e.g., laboratories combining several severe dangers.

In a similar manner, risk mapping is a technique used to help present identified risks and determine what actions should be taken towards those risks. Risk mapping is an approach to illustrating the risk associated with an organization, project or other system in a way that enables us to understand it better: what is important, what is not, and whether the risk picture is comprehensive. Risk mapping is primarily qualitative and its benefits are as follows:

- To improve our understanding of the risk profile and our ability to communicate about it.
- To force us to think through rigorously the nature and impact of the risks that have been identified.
- To improve our risk models by building an intermediate link between the risk register and the model.
- To improve our risk register by basing it on a more transparent and accurate understanding of the system.

A complete chapter on risk mapping would be out of scope of this book. Readers interested in having more insights are referred to (26), a book dealing with mapping wildfire hazards and risks.

2.6 Risk perception and risk attitude

Many decisions on risk taking, and other things in life, are colored by perception. The perception people have of risks is steered by a variety of factors that determine what risk is considered acceptable or unacceptable. Whether and how people expose themselves personally to certain risks (e.g., skiing, smoking, etc.), and how much, is a matter of preference and choice. The subjective factors on which, individuals base their decisions on to take a risk or not, include the degree to which the risk is known or unknown, voluntary or involuntary, acceptable or avoidable, threatening or attractive, controlled or uncontrolled.

Perception is very important, as if we have the mere perception that a risk is high, we will consciously or unconsciously take actions to reduce the risk. Also, where a person gives more weight to the positive side of a risk, he or she will be more prepared to take risks. Hence, influencing the perception is influencing the risk. Moreover, perception is fully linked with ERM, as the latter leads to an improvement of the perception of reality. By doing just that, ERM induces the taking of more objective decisions and of adequate actions.

Perception is strongly related with attitude. *Risk attitude* can be regarded as the chosen state of mind, mental view or disposition with regard to those uncertainties that could have a positive or a negative effect on achieving objectives. Hillson and Murray-Webster (27) explain that attitudes differ from personal characteristics in that they are situational responses rather than natural preferences or traits, and chosen attitudes may therefore differ depending on a range of different influences. If these influences can be identified and understood, obviously they can be changed and individuals and groups may than pro-actively manage and modify their attitudes. *Simply put, a person's risk attitude is his or her chosen response to perception of significant uncertainty.*

Different possible attitudes result in differing behaviors, which lead to consequences. This can be represented by the PABC-model:

Perception → **A**ttitude → **B**ehavior → **C**onsequences

As Hillson and Murray-Webster (27) indicate, although the responses to positive and negative situations suggest at first sight that situation is the foremost determinant of behavior, in fact it is how the situation is *perceived* by each person, because a situation that appears hostile to one may seem benign to another. This leads to the important question of what influences behavior when the situation is *uncertain*. In this case it is essential to know whether uncertainty is perceived as favorable, neutral, unfavorable, or hostile. This reaction to uncertainty is "risk attitude".

Risk attitudes exist on a spectrum. The same uncertain situation will elicit different preferred attitudes from different individuals or groups/organizations, depending on how they perceive the uncertainty. Hence, different people will behave differently to the same situation, as a result of their differing underlying risk attitudes and perceptions. The variety of possible responses to a given level of risk can be illustrated by ▶Fig. 2.11. The x-axis displays the comfort or discomfort level people relate to uncertainty. The y-axis displays the response that people will have on a given level of uncertainty. Different types of risk attitude can be seen in the figure, ranging from risk-paranoid to risk-addicted.

Obviously, the extreme attitudes (risk-paranoid and risk-addicted) are not common at all, and therefore the two well-known polarities are the risk-averse and risk-seeking attitude. On the one hand, a risk-averse person or group of persons feels uncomfortable with uncertainty, and therefore looks for certainty (safety and security) and resolution

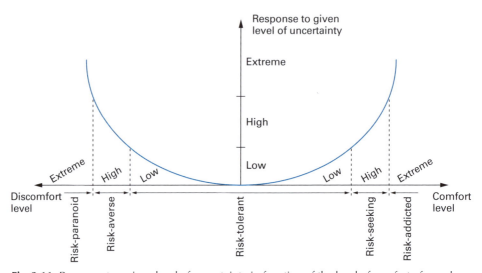

Fig. 2.11: Response to a given level of uncertainty in function of the level of comfort of people.

in the face of risks. Hazards and threats are perceived more readily and as more severe by the risk averse, leading to a preference for risk minimization and risk avoidance. Risk-seeking people on the other hand have no problem at all with uncertainty and therefore are likely to identify fewer threats and hazards, as they see these as part of normal business. Moreover, hazards and threats are likely to be underestimated both in probability and possible impact, and acceptance will be the preferred response. A risk-tolerant person or group of persons is situated somewhere in the middle of the spectrum and views uncertainty (and risk) as a normal part of life, and is reasonably comfortable with most uncertainty. This attitude may lead to a laissez-faire approach, which in turn may lead to a non-proactive mindset and thus non-proactive measures. As suggested by Hillson and Murray-Webster (2005) (27), this may be the most dangerous of all the risk attitudes. No proper risk management not only leads to important losses and problems, but also missed opportunities and missed hypothetical benefits (see also Chapter 9).

There are a number of situational factors that can modify the preferred risk attitude:

1. The level of relevant skills, knowledge or expertise: if high → risk-seeking attitude.
2. The perception of likelihood (probability of occurrence): if high → risk-averse attitude.
3. The perception of (negative) impact magnitude: if high → risk-averse attitude.
4. The degree of perceived control or choice in a situation: if high → risk-seeking attitude.
5. The closeness of risk in time: if high → risk-averse attitude.
6. The potential for direct (negative) consequences: if high → risk-averse attitude.

> The way in which people perceive reality is very important: increasing the accuracy of our perception of reality, leads to increasing our adequate ability to understand, to interpret, and to efficiently and effectively deal with risk.

2.7 ERM – main steps

Many flowcharts exist in the literature to describe the sequences of risk management. If we look through "engineering's goggles" we could draw the main steps involved in the ERM process. The process illustrated in ▶Fig. 2.12 is based on a structured and systematic approach covering all of the following phases: the definition of the problem and its context, risk evaluation, identification and examination of the risk management options, the choice of management strategy, intervention implementations, process evaluation and interventions, as well as risk communication. The phases are represented by circles, and the intersections show their interrelations.

The process normally starts at the problem definition step and proceeds clockwise. The central position of the risk communication phase indicates its integration into the whole process and the particular attention this aspect should receive during the realization of any of these phases.

Fig. 2.12: The engineering risk management process.

The process must be applied by taking into account the necessity of implementing the coordination and concentration mechanism, of adapting its intensity and its extent depending on the situation, and to enable revision of particular phases depending on the needs.

Although phases must generally be accomplished in a successive way, the circular form of this process indicates that it is iterative. This characteristic enables the revision of phases in light of all new significant information that would emerge during or at the end of the process and would enlighten the deliberations and anterior decisions. The made decisions should be, as often as possible, revisable and the adopted solutions should be reversible. Although the iterative character is an important quality of the process, it should not be an excuse to stop the process before implementing the interventions. Selecting an option and implementing it, should be realized even if the information is incomplete.

The flexibility must be maintained all along the process in order to adjust the relative importance given to the execution and the revision of the phases, as well as the depth level of analysis to perform or the elements to take into consideration.

The risk management process intensity should be adapted as best as possible to the different situations and can vary according to the context, the nature and the importance of the problem, the emergency of the situation, the controversy level, the expected health impact, the socio-economic stakes and the scientific data availability. As an example, emergency situations requiring a rapid intervention could require a brief application of certain phases of the process. However, even if the situation requires a rapid examination of these phases, all of them and all the indicated activities in this frame should be considered.

2.7 ERM – main steps | 25

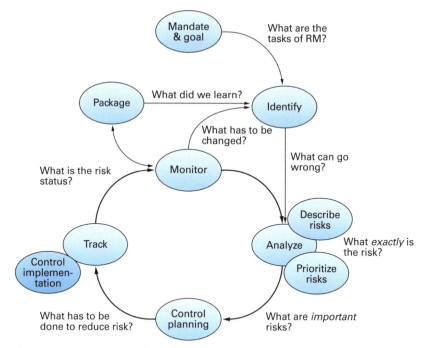

Fig. 2.13: Main questions of the ERM process.

Because this cycle has been similarly described many times in the literature, here we will go a little bit deeper. It is interesting to look at the risk management iterative ring through the questions that must be answered in order to get the process moving forward. A summary of these questions is presented in ▶Fig. 2.13. The starting point is the instruction or mission being the answer of, "What are the tasks of (negative) risk management?" Hence, we should identify "What could go wrong?" in the identification step. Answering "What exactly is the risk?" allows for describing, analyzing and prioritizing risks. Then, in order to control and plan, the question, "What are the important risks?" is raised. To implement the adequate measure for risk reduction, we have to answer, "What has to be done to reduce risk?" This allows also for controlling and tracking the implementation. The task is not yet over, as we should not forget to monitor the situation by asking several questions: "What is the risk status?" allows following the time evolution of the considered risk. If something begins to deviate, then, "What has to be changed?" brings us back to the risk identification step. Another important point, often forgotten in risk management, is the answer to "What did we learn?"

In summary, the ERM process is not only an identification and treatment process, it is a learning process that never ends and must be continuously performed.

Another characteristic of engineers is to simplify complex systems in order to master them more efficiently. From this perspective, we could imagine a simplification of the ERM process as depicted in ▶Fig. 2.14.

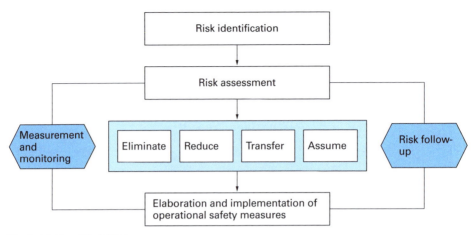

Fig. 2.14: Simplified ERM process.

Going back to the principles of risk management, ISO 31000:2009 (11) indicates that for risk management to be effective, an organization should at all levels comply with the principles below:

- Risk management creates and protects value. Risk management contributes to the demonstrable achievement of objectives and improvement of performance in, e.g., human health and safety, security, legal and regulatory compliance, public acceptance, environmental protection, product quality, project management, efficiency in operations, governance and reputation.
- Risk management is an integral part of all organizational processes. Risk management is not a stand-alone activity that is separate from the main activities and processes of the organization. Risk management is part of the responsibilities of management and an integral part of all organizational processes, including strategic planning and all project and change management processes.
- Risk management is part of decision-making. Risk management helps decision-makers make informed choices, prioritize actions and distinguish among alternative courses of action.
- Risk management explicitly addresses uncertainty. Risk management explicitly takes account of uncertainty, the nature of that uncertainty, and how it can be addressed.
- Risk management is systematic, structured and timely. A systematic, timely and structured approach to risk management contributes to efficiency and to consistent, comparable and reliable results.
- Risk management is based on the best available information. The inputs to the process of managing risk are based on information sources such as historical data, experience, stakeholder feedback, observation, forecasts and expert judgment. However, decision-makers should inform themselves of, and should take into account, any limitations of the data or modeling used or the possibility of divergence among experts.

- Risk management is tailored. Risk management is aligned with the organization's external and internal context and risk profile.
- Risk management takes human and cultural factors into account. Risk management recognizes the capabilities, perceptions and intentions of external and internal people that can facilitate or hinder achievement of the organization's objectives.
- Risk management is transparent and inclusive. Appropriate and timely involvement of stakeholders and, in particular, decision-makers at all levels of the organization, ensures that risk management remains relevant and up-to-date. Involvement also allows stakeholders to be properly represented and to have their views taken into account in determining risk criteria.
- Risk management is dynamic, iterative and responsive to change. Risk management continually senses and responds to change. As external and internal events occur, context and knowledge change, monitoring and review of risks take place, new risks emerge, some change, and others disappear.
- Risk management facilitates continual improvement of the organization. Organizations should develop and implement strategies to improve their risk management maturity alongside all other aspects of their organization.

The success of risk management will depend on the effectiveness of the management framework that provides the foundations and arrangements to embed it throughout the organization at all levels. The framework assists in managing risks effectively through the application of the risk management process at varying levels and within specific contexts of the organization. The framework ensures that information about risk derived from the risk management process is adequately reported and used as a basis for decision-making and accountability at all relevant organizational levels.

▶Fig. 2.15 shows the necessary components of the framework for managing risk and the way in which they interrelate in an iterative manner. We could observe that "communication and consultation" as well as "monitoring and review" are completely integrated in the ERM process. At almost all stages interrelations happen, the implementation of key factors is present (Key Performance Indicators or Key Risk Indicators) as the analysis of incident is crucial in the monitoring process.

Depending on the level of complexity necessary, we could use the different manners to represent what is ERM. Some users will prefer the simplified model (▶Fig. 2.14), while others will argue that the innovative representation with answering questions (▶Fig. 2.13) is preferable; some others will use the complete scheme (▶Fig. 2.15).

It is not really so important what scheme is used, the most important aspect is that with time one remains consistent in the use and in the follow-up. It is better to have a simplified system in adequate use rather than a complex scheme that will be only partially used.

> A variety of risk management schemes and frameworks are available to be used in industrial practice. A framework should always have a feedback loop built into it, where one is certain that risk management efforts never stop. Risk policy, assessment, communication, and monitoring should also always be part of the scheme.

2 Introduction to engineering and managing risks

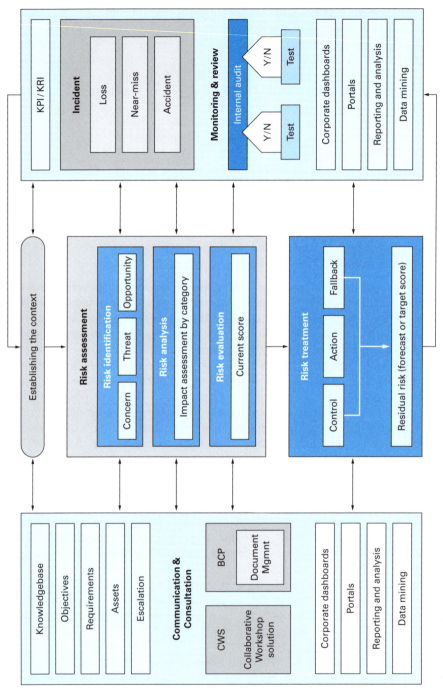

Fig. 2.15: Risk management process inspired from ISO 31000:2009.

2.8 Objectives and importance of ERM

Engineering today's organizations and their technology and systems is sophisticated and complex. Technology, processes, systems, etc. become ever more complex, distributed geographically and spatially, and they are linked through ever more complicated networks and subsystems. These operate to fulfill the needs of a variety of stakeholders, and they need to be ever more flexible, balanced in relation to expected performance, and risk managed in a responsible and respectful manner. ERM offers an answer to the ever more difficult task of managing the risks of organizational systems, technology, infrastructure, etc.

It is widely recognized that enacted law is by no means the only driver for improved risk management by corporations. Cost imperatives, liability issues, corporate reputation, industry peer pressure, etc., are increasingly important drivers. Community expectations about a better educated and informed workforce, more respect towards all stakeholders, and the absence of hazards possibly leading to major risks, enhance the use of ERM to achieve these expectations.

> We define the objectives of ERM as: "the early and continuous identification, assessment, and resolution of non-financial risks such that the most effective and efficient decisions can be taken to manage these risks."

There are many considerations explaining the importance of ERM, including early and continuous risk identification, risk-informed (risk-based) decision-making, gaining a systemic overview of an organization, proactive planning for unwanted events, proactive learning and learning from incidents, intelligent resource allocation, situational awareness and risk trends, etc.

ERM encompasses different well-known management disciplines such as change management (which is actually *ERM of changes*), project management (*ERM of projects*), crisis management (*ERM of crises*), innovation management (*ERM of innovations*), etc. ERM is only limited with respect to the type of operations, because only non-financial negative risks are envisioned with the term.

Employing the models, methodologies, ideas, theories, tools, etc. explained and elaborated in this book, leads to a systematic continuous improvement of organizational practices. This is shown in ▶Fig. 2.16.

The types of risk that can be handled by ERM, and by the methods described in this book, include accident risks, natural disasters, fire risks, technical risks, social risks, labor risks, etc. Handling these risks requires an analytical and systematic approach to identify important contributors to such risks, providing input in developing procedures for normal and emergency conditions, providing appropriate input to design processes, providing adequate inputs to the assessment of acceptability of potentially hazardous facilities, updating information on key risk contributors, deal with societal demands, etc.

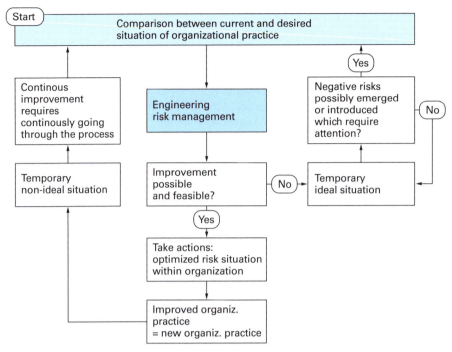

Fig. 2.16: Loop of continuous organizational practice improvement using ERM.

2.9 Conclusions

For having a thorough understanding of what constitutes *engineering risk management*, and how the theories, concepts, methods, etc. can be employed in industrial practice, some terms and concepts (uncertainty, hazard, risk, risk type, exposure, losses, risk perception, etc.) have to be unambiguously defined. In this chapter, we further focus on negative risks, and in what possible operational ways such risks can be decreased. The strategic part of engineering risk management is explained by presenting different possible risk management frameworks.

References

1. ISO 9001:2008 (2008) Quality Management Systems – Requirements. International Organization for Standardization. Geneva, Switzerland.
2. EFQM (2009) EFQM Excellence Model 2010, Excellent Organizations achieve and sustain superior levels of performance that meet or exceed the expectations of all their stakeholders. Brussels, Belgium: EFQM Publication.
3. ISO 9004:2009. (2009) Guidelines for Performance Improvement. International Organization for Standardization. Geneva, Switzerland.
4. ISO 14001:2004 (2004) Environmental Management Systems – Requirements with Guidance for Use. International Organization for Standardization. Geneva, Switzerland.

5. EMAS (Eco-Management and Audit Scheme) (2009) Regulation (EC) No 1221/2009 of the European Parliament and of the Council of 25 November 2009. Brussels, Belgium.
6. OHSAS 18001:2007 (2007) Occupational Health and Safety Management Systems – Requirements. London, UK: British Standardization Institute.
7. SA 8000:2008 (2008) Social Accountability Standard. Social Accountability International. New York.
8. Van Heuverswyn, K. (2009) Leven in de Risicomaatschappij Deel 1, Negen basisvereisten voor doeltreffend risicomanagement. Antwerpen: Maklu/Garant.
9. Government of Canada Treasury Board (2001) Integrated Risk Management Framework. Ottawa, Canada: Treasury Board of Canada.
10. AS/NZS 4360:2004 (2004) Australian-New Zealand Risk Management Standard. Sydney, Australia: Standards Australia International Ltd.
11. ISO 31000:2009 (2009) Risk Management Standard – Principles and Guidelines. International Organization for Standardization. Geneva, Switzerland.
12. Lees, F.P. (1996) Loss Prevention in the Process Industries. 2nd edn, vol. 3. Oxford, UK: Butterworth-Heinemann.
13. Wells, G. (1997) Major Hazards and their Management. Rugby, UK: Institution of Chemical Engineers.
14. Kletz, T. (1999) What Went Wrong? Case Histories of Process Plant Disasters. 4th edn. Houston, TX: Gulf Publishing Company.
15. Kletz, T. (2003) Still Going Wrong. Case Histories of Process Plant Disasters and How They Could Have Been Avoided. Burlington, VT: Butterworth-Heinemann.
16. Atherton, J., Gil, F. (2008) Incidents that Define Process Safety. New York: John Wiley & Sons.
17. Reniers, G.L.L. (2010) Multi-plant Safety and Security Management in the Chemical and Process Industries. Weinheim, Germany: Wiley-VCH.
18. Casal, J. (2008) Evaluation of the Effects and Consequences of Major Accidents in Industrial plants. Amsterdam, The Netherlands: Elsevier.
19. Balmert, P.D. (2010) Alive and Well at the End of the Day. The Supervisor's Guide to Managing Safety in Operations. Hoboken, NJ: John Wiley & Sons.
20. Rogers Commission report (1986). Report of the Presidential Commission on the Space Shuttle Challenger Accident.
21. Taleb, N.N. (2007) The Black Swan. The Impact of the Highly Improbable. New York: Random House.
22. US DoD (2002) Department of Defense, Office of the Assistant Secretary of Defense (Public Affairs). News Transcript, February 12.
23. Kirchsteiger, C. (1998) Absolute and relative ranking approaches for comparing and communicating industrial accidents. J. Hazard. Mater. 59:31–54.
24. Fuller, C.W., Vassie, L.H. (2004) Health and Safety Management. Principles and Best Practice, Essex, UK: Prentice Hall.
25. Marendaz, J.L., Suard, J.C., Meyer, Th. (2013) A systematic tool for Assessment and Classification of Hazards in Laboratories (ACHiL). Safety Sci. 53:168–176.
26. Sampson, R.N., Atkinson, R.D., Lewis, J.W. (2000) Mapping Wildfire Hazards and Risks. New York: CRC Press.
27. Hillson, D., Murray-Webster, R. (2005) Understanding and Managing Risk Attitude. Aldershot, UK: Gower Publishing Ltd.

3 Risk management principles

3.1 Introduction to risk management

The modern conception of risk is rooted in the Hindu-Arabic numbering system that reached the West some 800 years ago (1). Serious study began during the Renaissance in the 17th century. At that time, large parts of the world began to be discovered by the Europeans, and they began to break loose from the constraints of the past. It was a time of religious turmoil and a steep increase in the importance and rigorousness of mathematics and science. As the years passed, mathematicians transformed probability theory from a gamblers' toy into an instrument for organizing, interpreting and applying information. By calculating the probability of death with and without smallpox vaccination, Laplace developed the basis of modern quantitative analysis in the late 18th century.

The word "risk" seems to have been derived from the early Italian *risicare*, which means "to dare". In this sense, risk is a choice rather than a fate. The actions we dare to take, which depend on how free we are to make choices, are what "risk" should be about. The modern term "risk management" seems to have been first used in the early 1950s (2).

Whereas determining risk values as accurate as possible and obtaining a correct perception of all risks present (see Chapter 2) are both very important aspects of decreasing or avoiding the consequences of unwanted events, the aspect of making decisions about how to do this and what measures to take is at least as important. In contrast to risk assessment (including risk identification and analysis), which is a rather technical matter, risk management is largely based on company policy and can be seen as a response to perceptions. Therefore, risk management significantly differs across organizations, mainly as a result of the different values and attitudes towards specific risks in different organizational culture contexts.

As mentioned in Chapter 2, risk assessment connotes a systematic approach to organizing and analyzing scientific knowledge and information for potentially hazardous activities, machines, processes, materials, etc. that might pose risks under specified circumstances. The overall objective of risk assessment is to estimate the level of risk associated with adverse effects of one or more unwanted events from one or more hazardous sources. By doing so, it supports the ability to pro-actively and re-actively deal with minor accidents as well as major accidents through "risk management". Hence, risk-based decision-making consists of risk assessment and risk management. The former is the process by which the results of a risk analysis (i.e., risk estimates) are used to make decisions, either through relative ranking of risk reduction strategies or through comparison with risk targets; while the latter consists of the planning, organizing, leading and controlling of an organization's assets and activities in ways that minimize the adverse operational and financial effects of losses upon the organization. Risk assessment is thus only one element belonging to the larger domain of risk management. Other elements that are part of risk management are, e.g., safety training and education, training-on-the-job, management by walking around, emergency response, business continuity planning, risk communication, risk perception, psycho-social aspects of risk, emergency planning, etc. We may define "risk management" as: *the systematic application of management policies, procedures and practices to the tasks of identifying, analyzing, evaluating, treating and monitoring risks* (3). ▶Fig. 3.1 illustrates the risk management set.

A simplified overview of the risk management process to cope with all the tasks related to the risk management set, according to ISO 31000:2009 (4), is illustrated in ▶Fig. 3.2.

The sequence of the various process steps in assessing the risks originating from a specified system, i.e., establishing the context, identification, analysis, assessment, handling, monitoring, management and decision-making, is very similar at a generic level across different industries and countries. This observation also holds for different kinds of risk – safety-related risks, health-related risks, but also environmental risks, security

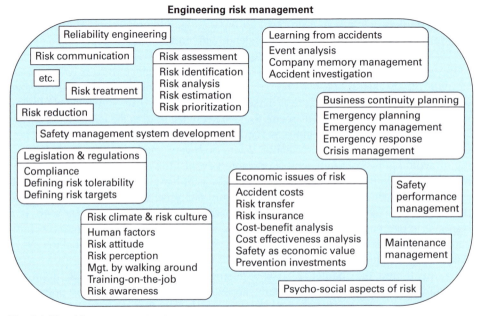

Fig. 3.1: The risk management set.

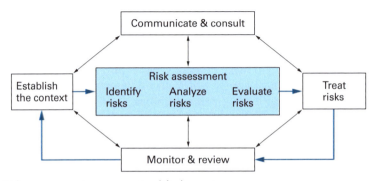

Fig. 3.2: Risk management process – simplified overview.

risks, quality risks, and ethical risks. The next section explains more in-depth how risk management should ideally be elaborated in a generic and integrated way.

3.2 Integrated risk management

When comparing a specific management system (designed for one domain, e.g., safety or environment or quality, etc.) with an integrated management system (designed for all risk domains), five common aspects can be found:

1. Both systems provide guidelines on how to develop management systems without explicitly prescribing the how-exercise in details.
2. They consider a risk management system to be an integral part of the overall company management system. This approach guarantees the focus and ability to realize the company's general and strategic objectives.
3. They all have two common aims: (i) to realize the organization's objectives taking compliance into full consideration; (ii) continuous improvement of an organization's achievements and performances.
4. The process approach is employed.
5. The models can be applied to every type of industrial organization.

The comparison also leads to the identification of several differences:

- Integrated risk management systems recognize the positive as well as the negative possible outcomes of risks. Hence, both damage and loss on the one hand, and opportunities and innovation on the other hand, are simultaneously considered.
- All kinds of risks are considered: operational, financial, strategic, juridical, etc. and hence, a balanced equilibrium is strived for.
- The objectives of integrated management systems surpass compliance and continuous improvement.

The question that needs to be answered, is how the integration of management systems can be realized in industrial practice, taking the similarities and differences of specific risk management systems into account. A suggested option is to integrate all management systems according to the scheme illustrated in ▶Fig. 3.3.

It is essential that an organization initially formulates a vision with respect to the system that is used to build the integrated risk management system. Such a vision demonstrates the intentions of the organization in respect to its risk management system in the long-term.

For a concrete elaboration of an integrated risk management system, the starting point can be the results of an organization (▶Fig. 3.4: Results) – its overall objectives, action plans and improvement measures. When these are known, instruments need to be designed for carrying out daily operations (▶Fig. 3.4: Approach) and realizing the established aims (▶Fig. 3.4: Implementation) [instruments can be, e.g., (working) procedures, (operational) instructions, standard forms, letters, etc.]. In a subsequent phase, instruments need to be figured out (▶Fig. 3.4: Evaluation) to assess the effectiveness and efficiency of goal realization within the company (e.g., employing risk analyses, threat assessments, SWOTs, internal and external audits, survey results, scoreboards, etc.). Finally, the results of the analysis on the different domains (safety, health, environment, ethics/integrity, security, quality, etc.), as well as documents of related meetings (e.g., safety, security, quality, management improvement, external audits, inspections, etc.), need to be

3 Risk management principles

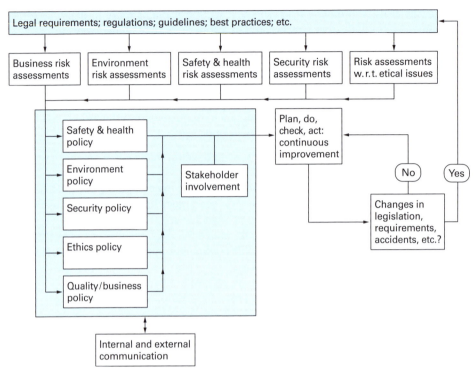

Fig. 3.3: Scheme of integrated risk management.

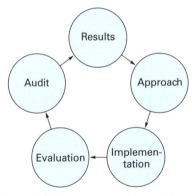

Fig. 3.4: Integrated risk management in practice: an approach.

used (▶Fig. 3.4: Audit) in follow-up meetings to define new objectives and improvement actions for the next time period. ▶Fig. 3.4 illustrates the entire integration process.

> Risk management requires an integrated view. Risks very often do not only have an impact on one domain, e.g., health and safety, but they also affect other domains such as environment or security, or quality. As such, the generic risk management process should be elaborated at – and implemented on – a helicopter level, for all domains at once.

3.3 Risk management models

Risk management models are described and discussed in this section, providing insights into organizational risk management. The models visualize the various factors and the different domains that the structure and approaches of adequate risk management should take into account within any organization.

3.3.1 Model of the accident pyramid

Heinrich (5), Bird (6) and Pearson (7), amongst other researchers, determined the existence of a ratio relationship between the numbers of incidents with no visible injury or damage, over those with property damage, those with minor injuries, and those with major injuries. This accident ratio relationship is known as "the accident pyramid" (see an example in ▶Fig. 3.5). Accident pyramids unambiguously indicate that accidents are "announced". Hence the importance of awareness and incident analyses.

Different ratios were found in different studies (varying from 1:300 to 1:600) depending on the industrial sector, the area of research, cultural aspects, etc. However, the existence of the accident pyramid has obviously been proven from a qualitative point of view. It is thus possible to prevent serious accidents by taking preventive measures aimed at near-misses, minor accidents, etc. These "classic" accident pyramids clearly provide an insight into type I accidents where a lot of data is at hand.

However, based on the different types of uncertainties/risks available, and based on several disasters (such as the BP Texas City Refinery disaster of 2005), the classic pyramid shape needs to be improved. Instead of an "Egyptian" pyramid shape, such as the classic accident pyramid that studies suggest, the shape should rather be a "Mayan" one, as illustrated in ▶Fig. 3.6.

The Mayan pyramid shape shows that there is a difference between type I risks and type II and III risks – in other words "regular accidents" (and the incidents going hand-in-hand with them) should not be confused with "major accidents". Not all near-misses have the potential to lead to disaster, but only a minority of unwanted events may actually eventually end up in a catastrophe. Obviously, to prevent disasters and catastrophes, risk management should be aimed at both types of risks, and certainly not only at the large majority of "regular" risks. Hopkins (8) illustrates this by using a bi-pyramid model, consisting of two pyramids

Fig. 3.5: The Bird accident pyramid ("Egyptian").

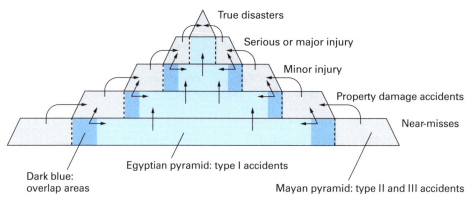

Fig. 3.6: Mayan accident pyramid shape.

partially overlapping. One pyramid represents type I risks, leading at most to a serious accident (e.g., a lethality), but not to a catastrophe, and the other pyramid represents type II and III risks, with the possibility to lead to a disaster. The overlap is also present in the Mayan pyramid of ▶Fig. 3.5, represented by the dark blue areas, because in some cases, unwanted events may be considered as warnings or incidents for both pyramids.

Thinking in the line of the classic Egyptian accident pyramid – resulting from the studies by Heinrich, Bird, Pearson and others – has had an important influence on dealing with safety in organizations, and it still has, but we should realize that this way of thinking is the cause as well as a symptom of *blindness towards disaster*. Therefore, the Mayan pyramid model or the bi-pyramid model are essential, and they provide a much better picture of how to deal with safety. Hopkins (8) indicates that the airline industry was the pioneer regarding this kind of safety thinking. In this industry, taking all necessary precaution measures to ensure flight safety is regarded as fundamentally different from taking prevention measures to guarantee employee safety and health. Two databases are maintained by airline companies: one database is used to keep data of near-miss incidents affecting flight safety, and another database is maintained to store information regarding workforce health and safety. Hence, in this particular industry we understand that workforce injury statistics tell nothing about the risk of an aircraft crash. This line of thinking should be implemented in every industrial sector.

In summary, the correct pyramid shape should be Mayan, making a distinction between the different types of risk. Type II and III risks do not only require constant individual mindfulness based on statistical modeling, but also collective mindfulness, focus on major accident, safety, an open mind towards extremely unlikely events, etc., based on qualitative and semi-quantitative risk approaches and assessments.

3.3.2 The P2T model

If accidents result from holes in the safety system, then it is important to "close the holes" in time. Based on the OGP model for Human Factors (9), Reniers and Dullaert (10) discern three dimensions in which measures can be taken to avoid and to prevent unwanted events, and to mitigate their consequences. The three dimensions are: *people, procedures and technology* (see ▶Fig. 3.7), and the model is therefore called the "P2T" model.

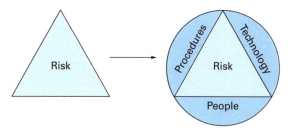

Fig. 3.7: People, procedures and pechnology to manage and control risks.

Applied to risk management within an organization, the first dimension, *people*, indicates how people (individually and in group) deal with risks and think about risks in the organization (including domains such as training, competence, behavior and attitude, etc.). The second dimension, *procedures*, concerns all management measures taken in the organization to tackle risks in all possible situations and under all conceivable circumstances (including topics such as work instructions, procedures, guidelines, etc.). The third component, *technology*, comprises all technological measures and solutions taken and implemented with respect to risk management (including risk software, safety instrumented functions, etc.).

3.3.3 The Swiss cheese model and the domino theory

The "Swiss cheese" model was developed by the British psychologist Reason (11), to explain the existence of accidents by the presence of "holes" in the risk management system (see ▶Fig. 3.8). A solid insight into the working of the organization allows for the possibility to detect such "holes", while risk assessment includes the identification of suitable measures to "close the holes". The cheese itself can be considered as the positive side of risks, and thus the more cheese, the more gains the risks may provide.

It is important to notice that the Swiss cheese is dynamic: holes may increase (e.g., caused by mistakes, errors, violations, lack of maintenance, etc.), but they may also decrease (because of solid risk management and adequate preventive and protective measures).

This model is very powerful in its use of "barrier" thinking (or "layer of protection" or "rings of protection" thinking). The holes within the barriers should be made as small as possible through adequate risk management, and this should be done for type I as well as type II and III risks. For type III risks, a further elaborated Swiss cheese based model can be suggested. ▶Fig. 3.9 illustrates the theoretical safety continuous amelioration picture, (especially interesting for preventing type II and III accidents, on top of type I accidents) – the P2T (people, procedures, technology) model from Section 3.3.2 is combined with the Swiss cheese model.

The upper half of ▶Fig. 3.9 shows the stepwise progress that risk management science has made over time and is still making. Taking two companies A and B as an example, every distinct progress step (characterized by new insights, and representing the dimensions from the P2T model) is represented by a rhomb. This rhomb can actually be considered as a kind of safety layer. The hypothetical development of five potential accidents for the two plants is shown. Every potential accident is prevented by one of the dimensions, elaborated on an individual plant level or on a multi-plant

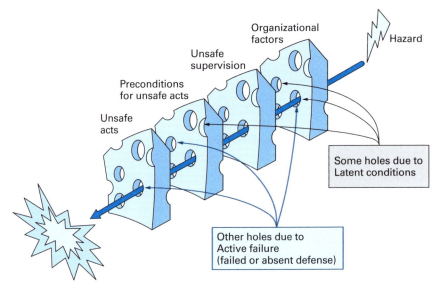

Fig. 3.8: The Swiss cheese model.

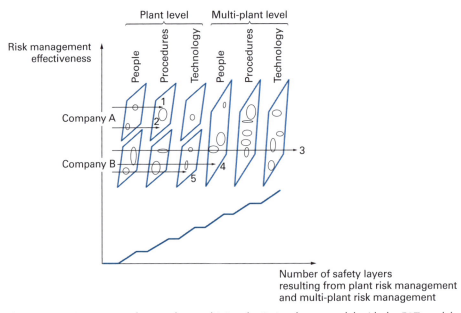

Fig. 3.9: Stopping unwanted events by combining the Swiss cheese model with the P2T model.

scale, except for accident number three. Accident number one, e.g., was stopped by the accident prevention layer marked by the existing procedures on a plant level. The lower half of the figure illustrates the needed ceaseless stepwise safety improvement resulting from new insights (captured over time) dealing with risks. Each safety layer (itself composed of a number of safety barriers) is considered as increasing safety effectiveness.

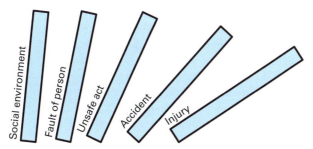

Fig. 3.10: The domino theory model. Source: Heinrich (12).

Linked with the Swiss cheese model, is the "domino model" by Heinrich (12–14) Heinrich, sometimes called the father of industrial safety and prevention, indicates that the occurrence of an accident invariably results from a completed sequence of factors, the last one of these being the accident itself. The accident further results in loss (injury, damage, etc.). The accident is invariably caused by the unsafe act of a person and/or a mechanical or physical hazard. ▶Fig. 3.10 illustrates the Heinrich domino theory model, and shows that the sequential accident model has a clear assumption about causality, specifically that there are identifiable cause-effect links that propagate the effects of an unwanted event.

An *accident can thus be visualized by a set of domino blocks* lined up in such a way that if one falls, it will knock down those that follow. This basic principle is used in accident investigations. It should be noted that the domino model by Heinrich should not be confused with other items such as "domino accidents" and "domino effects", indicating "chains of accidents" or escalating events whereby one accident leads to another accident, and so on.

Although the domino model by Heinrich only applies to type I risks and is somewhat antiquated [because present insights in risk management indicate that systems thinking (see Section 3.10.1) should be used, and not only a cause and effect model], it still has its merits for presenting a complex accident in a simple way.

3.4 The anatomy of an accident: SIFs and SILs

The best way to achieve safety within any organization is to have inherently safe processes, and to operate in correctly designed, controlled and maintained environments (from the perspective of people, procedures, and technology). However, it seems inevitable that once designs have been engineered to optimize organizational safety, a spectrum of risks remain in many operations. To deal with these risks, a comprehensive safety management system is developed. This safety management system addresses hazard assessment, specification of risk control measures, evaluation of the consequence of failures of these controls, documentation of engineering controls, scheduled maintenance to assure the on-going integrity of the protective equipment, etc. Safety instrumented functions (SIF[1]) can, e.g., be considered as a prevention measure. A SIF is a combination of sensors, logic solvers, and final elements with a specified safety

---

[1] This is a newer, more precise term for a safety interlock system (SIS).

integrity level (SIL) that detects an out-of-limit condition and brings a process to a functionally safe state. However, these methodologies do not investigate the underlying physical and chemical hazards that must be contained and controlled for the process to operate safely, and thus they do not integrate inherent safety into the process of achieving safe plant operation. Most action items refer to existing safety procedures or to technical safeguards, or require the addition of new levels of protection around the same underlying hazards. In other words, they represent "add-on safety". However, taking preventive measures in the conceptual design phase is extremely important to optimize and help achieving organizational safety.

The most desirable requirement of equipment is that it is *inherently safe*. Achieving such inherent safety starts in the design phase of the equipment. An inherently safe design approach includes the selection of the equipment itself, site selection and decisions on dangerous materials inventories and company layout. Complete inherent safety is rarely achievable within economic constraints. Therefore potential hazards remaining after applying such an approach should be addressed by further specifying independent protection layers to reduce the operating risks to an acceptable level.

In current industry practice, chemical facilities processing dangerous substances are designed with multiple layers of protection, each designed to prevent or mitigate an undesirable event. Multiple independent protection layers (IPL) addressing the same event are often necessary to achieve sufficiently high levels of certainty that protection will be available when needed. Powell (15) defines an IPL as having the following characteristics:

- Specific – designed to prevent or to mitigate specific, potentially hazardous events.
- Independent – independent of the other protective layers associated with the identified hazard.
- Dependable – can be counted on to operate in a prescribed manner with an acceptable reliability. Both random and systematic failure modes are addressed in the assessment of dependability.
- Auditable – designed to facilitate regular validation (including testing) and maintenance of the protective functions.
- Reducing – the likelihood of the identified hazardous event must be reduced by a factor of at least 100.

An IPL can thus be defined as a device, system or action that is capable of preventing a scenario from proceeding to its undesired consequence independent of the initiating event or the action of any other layer of protection associated with the scenario.

▶Fig. 3.11 illustrates safety layers of protection that are specifically used in the chemical industry. Detailed process design provides the first layer of protection. Next come the automatic regulation of the process heat and material flows and the providing of sufficient data for operator supervision, in the chemical industry together called the "basic process control systems" (BPCS). A further layer of protection is provided by a high-priority alarm system and instrumentation that facilitates operator-initiated corrective actions. A safety instrumented function, sometimes also called the "emergency shutdown system", may be provided as the fourth protective layer. The SIFs are protective systems that are only needed on those rare occasions when normal process controls are inadequate to keep the process within acceptable bounds. Any SIF will qualify as one IPL. Physical protection may be incorporated as the next layer of protection by

3.4 The anatomy of an accident: SIFs and SILs | 43

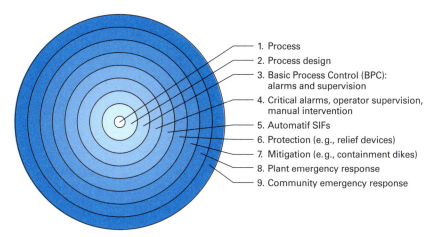

Fig. 3.11: Typical layers of protection found in modern chemical plants.

using venting devices to prevent equipment failure from overpressure. Should these IPL fail to function, walls or dikes may be present to contain liquid spills. Plant and community emergency response plans further address the hazardous event.

By considering the sequence of events that might lead to a potential accident (see also the "domino" model by Heinrich, in the previous subsection), another representation can be developed that highlights the efficiency of the protection layers, as shown in ▶Fig. 3.12 [adapted from (16)].

▶Fig. 3.12 illustrates the benefits of a hazard identification approach to examine inherent safety by considering the underlying hazards of a hazardous substance, a process or an operation. Inherently safer features in a design can reduce the required SIL (▶see Tab. 3.1) of the SIF, or can even eliminate the need for a SIF, thus reducing cost of installation and maintenance. Indeed, the Center for Chemical Process Safety (17) suggests that added-on barriers applied in non-inherently safer processing conditions have some major disadvantages, such as the barriers of being expensive to design, build and maintain, the hazard of being still present in the process, and the accumulated failures of IPLs still with the potential to result in an incident. An incident can be defined as "an undesired specific event, or sequence of events, that could have resulted in loss" whereas an accident can be defined as "an undesired specific event, or sequence of events, that has resulted in loss".

The primary purpose of risk evaluation methods currently employed is to determine whether there are sufficient layers of protection against an accident scenario. A scenario may require one or more protection layers depending on the process complexity and the potential severity of a consequence. Note that for a given scenario only one layer must work successfully for the consequence to be prevented (see also the Swiss cheese model and the Heinrich domino model). However, as no layer is perfectly effective, sufficient protection layers must be provided to render the risk of the accident tolerable. Therefore, it is very important that a consistent basis is provided for judging whether there are sufficient IPLs to control the safety risk of an accident for a given scenario. Especially in the design phase, an approach for drafting inherent safety into the design by implementing satisfactory IPLs is needed for effective safety. In many cases the SIFs are the final independent layers of

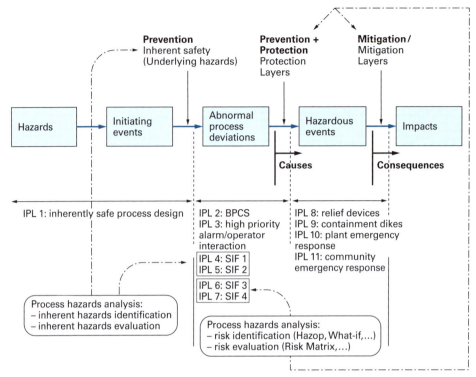

Fig. 3.12: Anatomy of an accident (with four SIFs).

Tab. 3.1: Safety integrity.

Safety integrity level (SIL)	Safety availability	Probability to fail on demand (PFD)	Equivalent risk reduction factor (1/PFD)
SIL 4	>99.99%	$\geq 10^{-5}$ to $<10^{-4}$	10,000–100,000
SIL 3	99.9–99.99%	$\geq 10^{-4}$ to $<10^{-3}$	1000–10,000
SIL 2	99–99.9%	$\geq 10^{-3}$ to $<10^{-2}$	100–1000
SIL 1	90–99%	$\geq 10^{-2}$ to $<10^{-1}$	10–100

Source: based on International Standard, 2003 (18).

protection for preventing hazardous events. Moreover, all SIFs are required to be designed such that they achieve a specified safety integrity level (SIL).

The SIL is the quantification of the probability of failure on demand (PFD) of a SIF into four discrete categories. The PFD can be defined as "the probability that a system will fail to perform a specified function on demand". ▶Tab. 3.1 gives an overview of such levels.

Thus, four corresponding degrees of reduction in hazardous event likelihood are produced by the SILs. SIL 1 provides about two orders of magnitude of event likelihood reduction; SIL 2 about three orders of magnitude, SIL 3 about four orders of magnitude and SIL 4 more than four orders of magnitude. Obviously, the availability targets for SIL 3

3.4 The anatomy of an accident: SIFs and SILs | 45

and SIL 4 are extremely stringent and the design practices to achieve and maintain these high levels are extensive and costly.

Gardner (19) points out that methods used to select safety integrity levels are based on an evaluation of three characteristics of the process and the hazardous event associated with the SIF: the severity of the hazardous event consequences (minor, serious, extensive), the likelihood that an upset situation will occur that could lead to these consequences (low, moderate, high) and the number of IPLs. Before a SIL can be selected, the inherent risk of the process must be evaluated. Next, credit for all non-SIF mitigation measures (e.g., relief valves, dikes) must be accounted for, to determine the baseline risk of the process, which is the starting point of the SIL selection. All of the SIF design, operation, and maintenance choices have to be verified against the target SIL. The safety design engineer has to realize that further mitigation with a SIF solely reduces the likelihood of an incident. For example, if the baseline likelihood is 10^{-2} per year, a SIL 2 would reduce the likelihood up to 10^{-5} per year. The risk reduction process is illustrated in ▶Fig. 3.13, in which risk criteria are represented in the form of *FN* limit lines (see also Section 3.5.2).

Determining the necessity of IPLs and SIFs, and the required level of their safety integrity, is performed using risk identification and evaluation methods. No one approach for the selection of a SIL is appropriate in every situation.

Summarizing, in highly technological environments, the SIL is chosen to reduce the incident frequency to a tolerable level. It is the design basis for all engineering decisions related to the SIF. When the design is complete, it must be validated against the SIL. Therefore, the SIL closes the design cycle: starting with hazards identification, then requirements quantification and ending with design validation.

Achieving inherent safety is not always possible. When processing, storing or transporting hazardous materials, rest risks often remain, to a greater or lesser extent. To

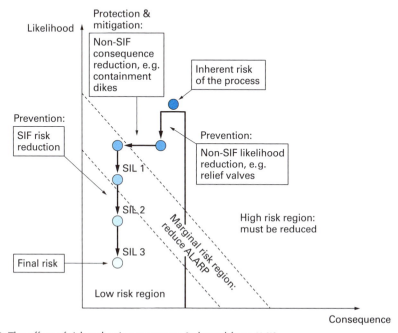

Fig. 3.13: The effect of risk reduction measures [adapted from (16)].

identify and to assess rest risks in an optimal and effective way, the analyses aimed at remaining risks should be thoroughly carried out.

The fundamental basis of security management can be expressed in a similar way to the layers of protection used in chemical process plants to illustrate safety barriers (see ▶Fig. 3.11). In the similar concept of concentric rings of protection (20), the spatial relationship between the location of the target asset and the location of the physical countermeasures is used as a guiding principle. ▶Fig. 3.14 [adapted from (16)] exemplifies the rings of protection and their component countermeasures, illustrating the responsibilities and the distinction between indoor and outdoor security guards.

In security terms, critical infrastructure is broadly defined as people (employees, visitors, contractors, nearby members of the community, etc.), information (formulae, prices, processes, substances, passwords, etc.), and property (buildings, vehicles, production equipment, storage tanks and process vessels, control systems, raw materials, finished products, hazardous materials, natural gas lines, rail lines, personal possessions, etc.) which are believed crucial to prevent major business disruption and resulting substantial economic and/or societal damage.

By considering the sequence of events that might lead to a potentially successful attack, another representation can be given, illustrating the effectiveness of the rings of protection (see ▶Fig. 3.15).

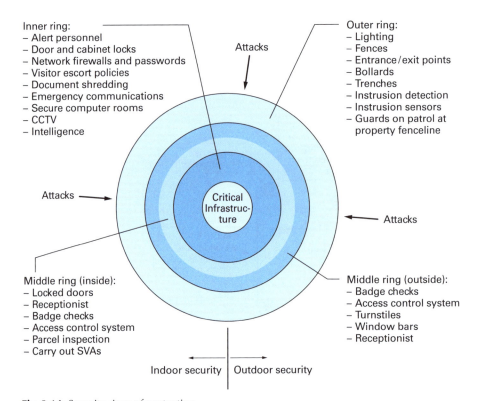

Fig. 3.14: Security rings of protection.

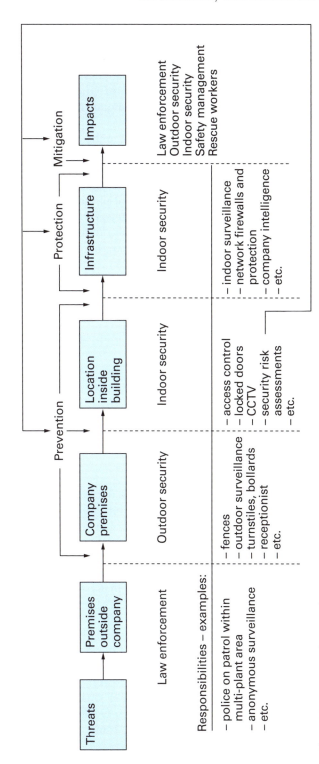

Fig. 3.15: Anatomy of an attack.

Firstly, companies can clearly protect themselves in a much better way against external attacks than against attacks from within the company itself, because in the latter case there only exists indoor security to avert the threat. Secondly, as the effective prevention, protection and mitigation of attacks depend on meticulously carrying out security risk assessments, the latter is of crucial importance to deter, detect and delay possible threats within a single company as well as within a cluster of companies.

Once the entire sequence of events has taken place, either safety-related or security-related, there is a loss. Regardless of the particular business activity in which the loss may have occurred, losses can be considered minor, serious, major, or catastrophic. Depending upon the particular business activity, the determining factors for rating the severity of an accident loss can be the degree of physical harm and/or property damage and any resulting humane or economic aspects. When evaluating either these humane or economic effects, investigators should be particularly aware that those factors readily apparent are usually indicative of much more serious and far-reaching aspects that are not so obvious. Much like the tip of an iceberg, the extent and size of the problems associated with accidents and losses are not easily seen or determined on the surface, but they are, nonetheless, there (21). We also refer to Chapter 8 on economic issues related to safety.

> Models used to deal with risks are very diverse. The models have been built after decades of experience and research, within a variety of academic disciplines, and encompassing diverse industrial sectors. Incidents and accidents were a driver and an inspiration for the builders of the models.

3.5 Individual risk, societal risk, physical description of risk

3.5.1 Individual risk

An individual risk can be defined in general as "the frequency with which a person may expect to sustain a specified level of harm as a result of an adverse event involving a specific hazard" (3). Hence, the individual risk can be used to express the general level of risk to an individual in the general population, or to an individual in a specified section of the community. There is widespread agreement in many countries that an additional risk from industrial activities of 10^{-6} per year to a person exposed to this risk, is a very low level compared with risks that are accepted every day. The reasoning is thus. On the one hand, the individual risk of getting killed from driving a car is estimated as 10^{-4} per year. If the level of (individual) risk is higher than driving a car, it is perceived as unacceptable. On the other hand, the individual risk of being struck by lightning is estimated as 10^{-7} per year, and this risk level is perceived to be so low that it can be accepted.

Legislators often use levels of individual risk with "specified level of harm" equal to "fatality" (such risks are called "individual fatality risks"), as a regulatory approach to setting risk criteria. In the process industries e.g., the individual (fatality) risk is defined as the risk that an unprotected individual would face from a facility if he/she remained fixed at one spot, 24 hours a day, 365 days per year.

Individual fatality risks are calculated by multiplying the consequences and the frequency. For example, if the severity of an industrial accident is such that there is a *p%* probability of killing a person at a specified location (the probability merely takes into account the level of lethality due to the accident (e.g., due to heat radiation or a pressure wave), but does not consider population figures), and the accident has a frequency of *f* per year, then the individual fatality risk at this particular location is *pxf* per year. Where there is a range of incidents that expose the person at that point to risk, the total individual fatality risk is determined by adding the risks of the separate incidents. Iso-risk contours – contours of which on each point the risk level is identical – can then be plotted around an industrial activity, and can be used to present the risk levels surrounding the activity. Sometimes different levels of individual fatality risk are defined to be allowed by the authorities for different types of location. For example, a distinction is made between industrial areas, commercial areas, parks and sport fields, schools, rest and nursing homes, hospitals, etc. Individual risks are often calculated using "quantitative risk assessment" (QRA; see Chapter 4, Section 4.9).

3.5.2 Societal risk

Calculating the individual fatality risk around a specific industrial activity does not make a distinction between the activity taking place, e.g., somewhere in the desert or within the center of a major city. The calculated individual risk will be the same, regardless of the number of people exposed to the activity. However, it is evident that in reality the level of risk will not be identical for both situations: the level of risk will be higher if the industrial activity would take place in the city, rather than in the desert. This is a result from people living in the neighbourhood of the activity, and thus being exposed to the danger. To take the exposed population figures into account, a "societal risk" is calculated. This is "the probability that a group of a certain size will be harmed – usually killed – simultaneously by the same event or accident" (22). It is presented in the form of an FN-curve. Each point on the line or curve represents the probability that the extent of the consequence is equal to or larger than the point value. These curves are found by sorting the accidents in descending order of severity, and then determining the cumulative frequency. As both the consequence and the cumulative frequency may span several orders of magnitude, the FN-curve is usually plotted on double logarithmic scales [see also (23)]. A log-log graph is obtained, as depicted in ▶Fig. 3.16.

The societal risk is designed to display how risks vary with changing levels of severity. For example, a hazard may have an acceptable level of risk for just one fatality, but may be at an unacceptable level for 100 fatalities. In some jurisdictions, there are rigidly defined boundaries on the societal risk graph between the zones of high, intermediate, and low risk (see ▶Fig. 3.17). It is common, where an industrial activity is calculated to generate risks in the intermediate zone between high and low, to require the risks to be reduced to a level that is "as low as reasonably practicable" (ALARP), provided that the benefits of the activity that produces the risks are seen to outweigh the generated risks.

As the severity of the event increases, people become more risk averse. Particularly, once the death threshold is passed, it appears the community has a much greater

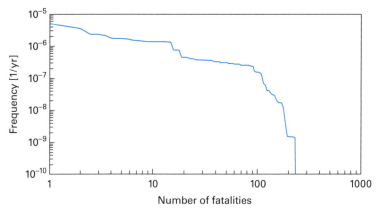

Fig. 3.16: FN curve – an illustrative example.

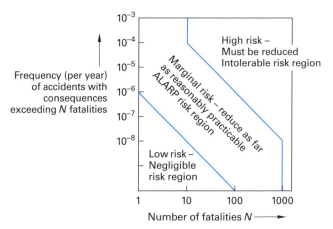

Fig. 3.17: Example of approach to defining societal risk criteria.

aversion to multiple fatality accidents. In many countries this seems to amount to a 100-fold decrease in the likelihood for the event for a 10-fold increase in the severity of the consequences measured in fatalities. This is shown in the FN-curves defined by the authorities from different countries to indicate the region where societal risks must be reduced and where they can be tolerated, depicted in ▶Fig. 3.18.

One measure of the societal risk from an installation could be obtained by calculating the fatal accident rate (FAR) of the number of fatalities per year from accidents involving dangerous substances. If the *FN-curve* of the installation is known then the value of FAR can be calculated as follows:

$$\text{FAR} = \sum_{N=1}^{N_{max}} f(N) \cdot N$$

However, using the FAR as a criterion for societal risk attracts criticism, for it does not include an allowance for aversion to multi-fatality accidents. It gives equal weight

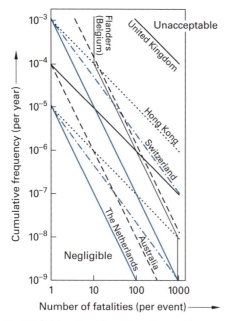

Fig. 3.18: Societal risk criteria in different countries.

to the frequencies and consequences of accidents. By not distinguishing between one accident causing 50 fatalities and 50 accidents each causing one fatality over the same period of time, the *FAR* fails to reflect the importance society attaches to major accidents (24). Moreover, several industrial activities from nearby companies in the same industrial area may each generate a low level of societal risk, whereas their combined societal risk might fall within the high-risk zone of the chart of ▶Fig. 3.17 (or ▶Fig. 3.18), if these industrial activities were all to be grouped for the purposes of the calculation.

In enterprises encountering numerous hazards with severe potential consequences, computer programs are used to calculate the risk levels on a topological grid, after which they are used to plot contours of risk on the grid. These contours are used to display the frequency of exceeding excessive levels of hazardous exposure. For example, Reniers et al. (25) discuss software tools that are available that will prepare contours, e.g., for the frequency of exposure to nominated levels of heat radiation, explosion overpressure and toxic gas concentration. Quantitative risk assessment (QRA) software is employed to plot *FN*-curves.

We should keep in mind that all these software tools provide insight into the possible scale of a disaster and the possibility of its occurrence, but they do not offer adequate information for the *optimal prevention* of catastrophic accidents. Moreover, if the number of events observed in the past is not sufficient to estimate significant frequency values, as in the case of type II and III events, a simple histogram plotting the absolute number of past events versus a certain type of consequence is often used instead of a risk curve. However, type II and III event predictions are extremely difficult to make simply because of the lack of sufficient data. Although it is thus not possible to take highly specific precaution measures based on statistic predictive information in such cases, engineering risk management, leading to a better understanding of relative risk levels and to an insight in possible accident and disaster scenarios, is actually essential to prevent such disastrous accidents and therefore should in one way or another be fully incorporated in industrial activities worldwide.

> In summary, there are different possible ways to calculate risk. Two well-known approaches, widely used, are the calculation of the individual risk and of the societal risk. An individual risk provides an idea of the hazardousness of an industrial activity. A societal risk takes the exposure of population in the vicinity of the hazardous activity into account in the calculation.

3.5.3 Physical description of risk

In the previous chapter, we defined a negative risk and indicated that all such risks are characterized by three factors: hazards – exposure – losses, together forming the "Risk Trias". Risk is a theoretical concept, and can be described in yet another way. To have a profound understanding of risk, we also discuss this second – more physical – approach.

In order to physically describe what a risk is, we must define some of its key components. The notion of the "target" needs to be introduced. By definition the target can be represented by:

- A human
- The environment
- A natural monument
- A process in a company
- A company
- The brand image, etc.

A threat is the potential of a hazard to cause damage. A threat can be intentional – then it is always a human threat related to the field of security – or it can be accidental or by coincidence. A threat is the direct consequence that arises by its link with the hazard; if the threat is non-intentional, it is subordinated to the law of probability.

Risk exists as soon as a hazard affects one or many possible targets. An identified hazard that does not affect any target does not represent a risk. For example, life on Mars may be very hazardous, but as long as nobody lives on Mars ... there are no losses, and hence no risk. Risk is found at the interface, or at the cross section, of a hazard and a target, as illustrated in ▶Fig. 3.19.

Basically, a risk is physically characterized by four elements:

1. A hazard.
2. One or many targets threatened by the hazard.
3. The evaluation of the threat.
4. The measures taken to reduce the threat.

3.5 Individual risk, societal risk, physical description of risk | 53

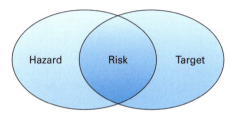

Fig. 3.19: Physical risk model.

These elements, depicted in ▶Fig. 3.20, show that a protection and/or prevention barrier is required in order to prevent a threat reaching the target.

The main difference between an incident and an accident is generally defined by the importance of caused or sustained damage (▶Fig. 3.21).

We defined incident and accident in Section 3.4. To have a more physical description of these terms, the norm OHSAS 18001 (26) can be used. It defines the notions of accident as follows:

- Incident: an event that leads or could have led to an accident.
- A near-accident or near-miss is an incident that does not damage health or leads to any deterioration or losses.
- Accident: an unexpected event that leads to health deterioration, lesions, damages or other losses.

A disaster is a major accident. It is an event that is brutal and sudden and of an enormous dimension. It has severe consequences that are accompanied by destruction of goods and/or death.

There are two possibilities to physically describe the risk: static and dynamic modeling.

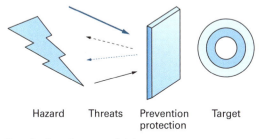

Fig. 3.20: Constitutive elements of risk.

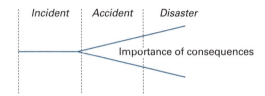

Fig. 3.21: Relation between incident, accident and disaster.

3.5.3.1 Static model of an accident

If risk is a potential, the accident is reality. It is realized as soon as a threat gets in contact with a target, allowing damage creation (▶Fig. 3.22). The prevention or protection barrier has only partially done its job as protector. The failure of the barrier is represented by the holes in the wall. For a static model the time scale is not included.

3.5.3.2 Dynamic model of an accident

While risk is not an event, the accident is one. Often it is the sequence of other events, the succession of an incident, which leads to damage. These successive incidents induce situations that are more and more hazardous, in which the likelihood of occurrence increases to a critical level and finally overcomes, as illustrated in ▶Fig. 3.23.

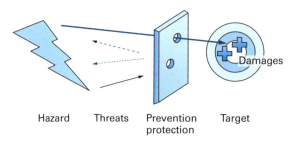

Fig. 3.22: Static model of an accident.

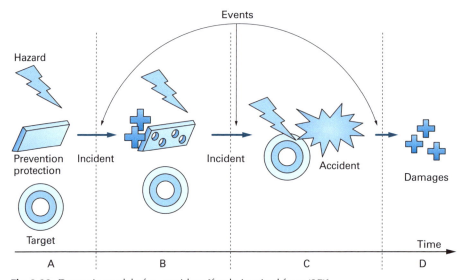

Fig. 3.23: Dynamic model of an accident [freely inspired from (27)].

We observe four zones:

A. At time zero, the protective/preventive barrier is fulfilling its purpose and prevents any threats to reach the target.
B. Small incidents happened, with time decreasing the protective/preventive level or efficiency of the barrier. It is the first observable sign of precursors for a future accident.
C. As time continues, the degradation of the barrier is now sufficient for the hazard (its threats) to reach the target. We now speak about an accident!
D. Finally the consequences of the accident are losses and damages.

The accident can also be the result of a situation that has changed continuously without the safety barrier measures being adapted to the changed situation. This is the case in many companies: the established measures have not been increased even if the company grew during a certain period. The protective and preventive measures are not adequate for the new size, and therefore risk becomes more and more important (▶Fig. 3.24).

Again four zones are observed:

A. At time zero the protective/preventive barrier is fulfilling its purpose and prevents any threats from reaching the target.
B. The target begins to grow with time (it is rarely that the barrier shrinks with time) but the barrier is still sufficient.
C. As time continues, the actual target is now larger than the size of target against which the barrier was designed to protect. We now speak about an accident!
D. Finally, the consequences of the accident are losses and damages.

Dynamic modeling taught us that:

- An accident is often the conclusion of successive events or incidents occurring subsequently, whether there is damage or not. Do not trivialize the incident!

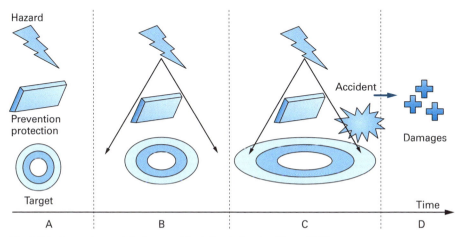

Fig. 3.24: Dynamic model of an accident [freely inspired from (27)].

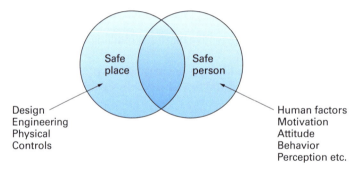

Fig. 3.25: Safe situation according to the classical risk modeling.

- The accident may be the result of a situation that has gradually changed without the provisions put in place originally being re-evaluated. Adapt measures to situations that have evolved.

To fulfill its protective role, risk management must be a dynamic, proactive and reactive process, based on careful and continuous observation of risk situations.

Finally, risk is not only a matter of technology but mainly dealing with humans. Therefore in order to get a safe situation, one must combine the safe place AND the safe person as illustrated in ▶Fig. 3.25.

> A risk was theoretically characterized in the previous chapter as hazard, exposure and loss. Risk management aims to decrease one of these elements, or a combination thereof. When physically describing a risk, it can be characterized as hazard, threat, prevention/protection, and target. In industrial practice, risk management aims to identify all of these, and by manipulating one of these elements, or a combination thereof, risks are managed.

3.6 Safety culture and safety climate

3.6.1 Organizational culture and climate

An organizational culture is: "a pattern of shared basic assumptions learned by a group as it solves its problems of external adaptation and internal integration, which has worked well enough to be considered valid and, therefore, to be taught to new members as the correct way to perceive, think, and feel in relation to those problems" (28). An organizational culture has an impact on the behavior of the employees, the operations and the results of the organization. In order for this impact to be positive, it is important that there is a good fit between the strategy and the culture of the organization (29). However, a company's culture will not be the only factor influencing the achievements and the excellence of a company. Two other influential factors are the organization's structure and its processes (30). For an extensive overview of definitions for "organizational safety culture", see Guldenmund (31).

Organizational culture can be analyzed at several different degrees or levels to which the cultural phenomenon is visible to the observer (28). The three levels are:

1. Artifacts (visible structures and processes as well as observable behaviors).
2. Espoused beliefs and values (ideals, values, aspirations, ideologies, rationalizations).
3. Basic underlying assumptions (unconscious, taken-for-granted beliefs and values).

It is essential to realize that a correct interpretation of the (most visible) artifacts depends on the knowledge of the pattern of basic assumptions. Although the essence of a group's culture is its pattern of shared, basic taken-for-granted assumptions, the culture will manifest itself at the level of observable artifacts and shared espoused values, norms and rules of behavior.

In addition to a company's culture, another important concept is a company's climate [e.g., see (32–36)]. Although both concepts are closely linked, it is imperative to make a clear distinction. Generally speaking, a company's climate can be thought of as "the product of some of the underlying assumptions and hence, it is the way in which a company's culture is visible to the outside world". Therefore, a company's climate can be seen as the outer layers of a company's culture and actually the manifestation of the culture. As a result, a company's culture emphasizes continuity, while its climate is comparable to a snapshot of its culture. An important difference between these two concepts is the way in which they are measured. A company's climate corresponds to the outer and more visible layers of its culture and can therefore be measured with objective, quantitative and qualitative methods, e.g., standardized questionnaires. A company's culture is more fundamental, and is for that reason more difficult to measure. Solely qualitative methods can be used, e.g., in-depth interviews (37). In any case, both an organization's culture and its climate should be integrated into a single model to truly advance an organizational domain through the process of goal-setting, measuring and continuously improving. To achieve excellence in a particular domain (such as safety and security e.g.), an organization needs to set goals and measure if it has reached those goals, i.e., it has to manage its performances in the domain. Organizational consensus must be achieved on what to measure, how to measure it, and what to do when corrections are needed. In order to determine what to measure, firstly the most important dimensions and sub-dimensions of a company's culture and climate need to be identified. A dimension or sub-dimension is a part of the company that is of critical importance to realize the corporate mission and strategy. Next, these dimensions and sub-dimensions should be translated into measurable performance indicators to be able to objectively measure them (38). If the company's track record is found to be inadequate in whatever domain, the organizational culture and/or climate needs to be changed or adapted by taking corrective measures and/or by setting new corporate objectives in the domain.

3.6.2 Safety culture models

The safety culture and its policy are defined as "the core values and behaviors resulting from a collective commitment by leaders and individuals to emphasize safety over competing goals to ensure protection of people and the environment". It should emphasize the importance of fostering and maintaining an open, collaborative work environment that encourages all employees and contractors to promptly speak up and share concerns and differing views without fear of negative consequences. The safety culture of an organization is the product of individual and group values, attitudes, perceptions, competencies and patterns of behavior that determine the commitment to, and the style and proficiency of, an organization's health and safety management.

There is no single definition of "a safety culture". The term first arose after the investigation of the Chernobyl nuclear disaster in 1986, which led to safety culture being defined as an organizational atmosphere where safety and health is understood to be, and is accepted as, the number one priority. In high-risk industries like aviation, nuclear power, chemical manufacturing, and fuel transportation, this makes sense. However, the problem is that safety and health does not exist in a vacuum isolated from other aspects of organizations such as people and financial management, as it both influences and is influenced by them, so safety culture is really a part of the overall corporate culture. On this basis, a more realistic definition may be "a safety culture is an organizational atmosphere where safety and health is understood to be, and is accepted as, a high priority". Some indicators for safety culture could be expressed as:

- Commitment at all levels.
- Safety and health are treated as an investment, not a cost.
- Safety and health is part of continuous improvement.
- Training and information is provided for everyone.
- A system for workplace analysis and hazard prevention and control is in place.
- The environment in which people work is blame free.
- The organization celebrates successes.

To develop a safety culture, change needs to be driven from the highest levels. The extent to which you can influence the organization largely depends on your place within the hierarchy. The recognition of the importance of a safety culture in preventing accidents has led to a growing number of studies to define and assess safety culture in a variety of complex, high-risk, industries.

An unsafe culture is more likely to be involved in the causation of organizational rather than individual accidents. Safety cultures evolve gradually in response to local conditions, past events, the character of the leadership and the mood of the workforce. An ideal safety culture is the "engine" that drives the system towards the goal of sustaining the maximum resistance towards its operational hazards, regardless of the leadership's personality or current commercial concerns.

According to Reason (39), several powerful factors act to push safety into the background of an organization's collective awareness, particularly if it possesses many elaborate barriers and safeguards. But it is just these defenses-in-depth that render such systems especially vulnerable to adverse cultural influences. Organizations are also prey to external forces that make them either forget to be afraid, or even worse, avoid fear altogether. The penalties of such complacency can be seen in the recurrent accident patterns in which the same cultural drivers, along with the same uncorrected local traps, cause the same bad events to happen again and again.

An organization with a "safety culture" is one that gives appropriate priority to safety and realizes that safety has to be managed like other areas of the business. That culture is more than merely avoiding accidents or even reducing the number of accidents, although these are likely to be the most apparent measures of success. It is to do the right thing at the right time in response to normal and emergency situations. The quality and effectiveness of that training will play a significant part in determining the attitude and performance – the professionalism. And the attitude adopted will, in turn, be shaped to a large degree by the "culture" of the company.

The key to achieving that safety culture is as follows:

- Recognizing that accidents are preventable through following correct procedures and established best practice.
- Constantly thinking about safety.
- Seeking continuous improvement.

It is relatively unusual for new types of accidents to occur, and many of those that continue to occur are caused by unsafe acts. These errors, or more often violations of good practice or established rules, can be readily avoided. Those who make them are often well aware of the errors of their ways. They may have taken short-cuts they should not have taken. Most will have received training aimed at preventing them but, through a culture that is tolerant to the "calculated risk", they still occur.

Following the Chernobyl accident in 1986, a lot more attention was paid to the term "corporate safety culture" and various definitions were proposed. Cooper (40) defines corporate safety culture as "that observable degree of effort by which all organizational members direct their attention and actions toward improving safety on a daily basis", hereby stressing that in a good safety culture all members of an organization should deliver intentional efforts to continuously improve overall safety. Hale (41) refers to beliefs, values and perceptions of natural groups within an organization and the effect of these groups on values and norms. These values and norms will define how a company will handle its risk and risk control systems. Wiegmann et al. (42) try to capture all previous definitions of safety culture in the following elaborated formulation:

> "Safety culture is the enduring value and priority placed on worker and public safety by everyone in every group at every level of an organization. It refers to the extent to which individuals and groups will commit to personal responsibility for safety, act to preserve, enhance and communicate safety concerns, strive to actively learn, adapt and modify (both individual and organizational) behavior based on lessons learned from mistakes, and be rewarded in a manner consistent with these values."

It is important to also notice that the aspect of learning from mistakes is adopted in the latter definition. Mohamed (43) presents a very pragmatic definition by stating that a corporate safety culture is a mere subculture of the general organizational culture. He defines safety culture as: *"a sub-facet of organizational culture, which affects workers' attitudes and behavior in relation to an organization's on-going safety performance"*. In summary, a safety culture can be regarded and explained as the way that people behave and act (with respect to safety) in an organization when nobody is watching them.

Moreover, as already mentioned in the previous section, an important difference exists between *safety culture* and *safety climate*. Much like organizational culture and climate, Hale (41) states that safety culture is the whole of values and practices that are linked to the company in a strong, unobservable relation. These values and practices are stable in time and cannot easily be observed nor be altered. The corporate safety climate is easier to observe. The explicit artifacts and values of a company are good examples of the safety climate components. Wiegmann et al. (42) add to this that the safety climate is a snapshot of the safety culture at one moment in time, and that the climate displays what is the perception of the culture by the members of an organization.

Safety culture is commonly viewed as an enduring characteristic of an organization that is reflected in its consistent way of dealing with safety issues. Safety climate is viewed as a temporary state of an organization that is subject to change depending on the features of the specific operational or economic circumstances. Therefore, just like personality researchers, safety researchers have attempted to identify key indicators of organizational safety culture and to develop methods for assessing the extent to which these key organizational features are consistent across time and situations.

Safety climate is the temporal state measure of safety culture, subject to commonalities among individual perceptions of the organization. It is therefore situationally based, refers to the perceived state of safety at a particular place at a particular time and is relatively unstable and subject to change, depending on the features of the current environment or prevailing conditions.

Safety climate is a psychological phenomenon, which is usually defined as the perceptions of the state of safety at a particular time. It is closely concerned with intangible issues such as situational and environmental factors. Hence, safety climate is a temporal phenomenon, a "snapshot" of safety culture, relatively unstable and subject to change.

Safety behavior presents a paradox to practitioners and researchers alike because, contrary to the assumption that self-preservation overrides other motives, careless behavior prevails during many routine jobs, making safe behavior an on-going managerial challenge.

At present there is no overall satisfying model for safety culture and climate, or for security[2] culture and climate (44). However, important features and capabilities that are essential for characterizing, elaborating and improving an organization's safety and security culture and its safety and security climate have been put forward by various authors. Moreover, as Guldenmund (31) rightly puts it, when a given safety and security culture and climate has been assessed, the next question will certainly be – so what? The organization needs to make conclusions, and corrective actions have to be taken and carried out if required. The next section presents a model that may be viewed and used as an easy-to-use approach to continuously improve the organization's safety and security culture or climate in order to achieve safety and security excellence and leadership.

3.6.3 The P2T model revisited and applied to safety and security culture and climate

In Section 3.3.2, we proposed a three-dimensional model that can be applied to develop an integrative safety and security culture and climate. With this model, all safety and security culture and climate aspects can be integrated and covered, as all elements concerning a good safety and security culture and climate can be placed under one of the three dimensions. The suggested dimensions were people, procedures and

---

[2] Security is characterized with intentionality and indicates deliberate acts by humans to cause loss. Safety indicates accidental loss.

3.6 Safety culture and safety climate

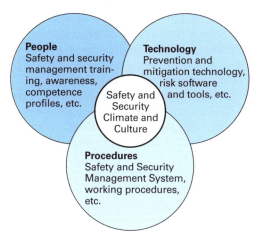

Fig. 3.26: Safety and security culture and climate according to the P2T-model.

technology, and the model was therefore referred to as the P2T-model. The interplay between the three domains defines the present safety and security culture and climate in any organization; see ▶Fig. 3.26.

We need little argument that the technological dimension is indispensable to ensure a good safety and security culture and climate. With failing installations or equipment, a company inflicts a direct threat to its workers and surrounding people and buildings. Despite the large improvements in safety technology that have been gained in the past decades, security technology such as CCTV and biometric systems e.g., have not been applied to their full potential in many enterprises. Following the ALARA-principle (as low as reasonably achievable), it should be noted that no risk can be reduced to zero without expenses that can be justified economically. That is why the technological dimension has to be designed in a way that the resulting risk lays between socially accepted boundaries. Governments will impose these bounds, but often organizations will surpass the required measures.

The second dimension, procedures, is being managed by a safety management system (in case of safety), and a security management program (in case of security). These management systems revise the existing procedures used to maintain a good safety culture and/or security culture. The term "procedures" can be interpreted very broadly. It concerns procedures to operate safely and securely; to safely store hazardous substances; to manage the competences of employees; to manage emergency situations; to have deter, detect, delay procedures in place, etc. Logically, the organizational structure and culture play a large role in this.

The third dimension to influence safety and security culture and climate is being defined as people. Reason (11), Fuller and Vassie (3), CCPS (45) and many other researchers indicate that a majority of accidents and near-misses can be attributed to human error. According to some estimates, human error contributes to 90% of all accidents (46). This number considers all possible sources of error, including front-line operating personnel, engineers, and supervision. In case of security, even all incidents are human made. Another point for the security is the necessity to understand the human element for threat and vulnerability analysis. Also, human errors can be made in analyzing, identifying, and responding to security incidents and this must be considered to minimize the threats and decrease risk. That is why creating safety and security awareness among all employees is essential for a good safety and security culture and climate, as well as providing proper training, providing safety and security incentives, creating a safety-driven and security-driven organizational community, enhancing competences of employees at all levels, etc.

Every dimension (people, procedures and technology) can be looked upon from the three cultural levels explained by Schein (28): artifacts, espoused beliefs and values, and basic underlying assumptions. This way, these three dimensions can be used to measure the safety culture (long-term objectives and results) and safety climate (short-term objectives and results).

It is evident that a good safety culture and climate depend on adequate and solid strategic management concerning risks. Section 3.7 discusses how strategic management should lead to continuous improvement, and Section 3.8 proposes and discusses a model to unify the principles of performance management and continuous improvement with the concepts of safety culture and climate.

3.7 Strategic management concerning risks and continuous improvement

Strategic management consists of five phases.

- The first phase determines the strategic vision of the organization. This vision makes it clear for the entire organization how the organization should look like and how it should evolve.
- The second phase consists of translating the strategic vision into clearly measurable objectives. This enables the company to measure if the desirable results have been achieved.
- In the third phase, the organization should develop a strategy that makes it possible for the organization to reach its goals. This strategy should be specified for each functional domain within the organization, e.g., the safety domain.
- In the fourth phase, the chosen strategy should be implemented in an efficient and effective way.
- Finally, in the fifth phase, the performance of the organization should be evaluated and if necessary, changes should be implemented (47).

In order to execute the last phase of strategic management a company should have a system of continuous improvement. One of the most widely used systems is the well-known *plan-do-check-act* loop of continuous improvement or the *Deming cycle*. The different steps of the Deming loop, which is also called the PDCA cycle, are below:

- Plan – develop a policy and determine the goals and processes necessary for achieving certain objectives, based on the risk analyses carried out and the subsequent action programs.
- Do – execute the actions and measures and realize the policy objectives; implement the processes (e.g., the risk management process).
- Check – monitor, measure and analyze the realization of the aims, targets, action programs, etc. and their effects, by means of inspections, audits, etc.; The result

Fig. 3.27: Basic philosophy of PDCA applied to risk management.

of these measurements and analyses is the definition of new corrective and/or preventive improvement actions, that aim to improve organizational processes and products in relation to the policy and its objectives.
- Act – take measures to continuously improve process achievements; review and eventually revise company policy.

The Deming cycle describes the quality management principle of continuous improvement that needs to be applied to all aspects, processes and activities throughout the whole company (44). The PDCA cycle can be illustrated in different ways and it can be filled in for different types of activities. ▶Fig. 3.27 illustrates the basic philosophy of the PDCA cycle for risk management.

3.8 The IDEAL S&S model

In current industrial settings, companies follow the *plan-do-check-act* loop of continuous improvement because of their acquired know-how of internationally accepted business standards such as the ISO9000 series and the ISO14000 series, addressing quality and environment management systems respectively. The OHSAS18000 series – the international Occupational Health and Safety management Assessment System specification that empowers an organization to control its occupational health and safety risks and improve its performance concerning those risks – also uses the PDCA cycle as a basic management concept and is often used to work out safety management systems.

3 Risk management principles

The ISO norms and OHSAS are very well-known throughout all industrial sectors, and hence some degree of basic management standardization often already exists. Available risk management models mention performance management as a technique to measure safety performance using proactive/leading indicators or reactive/lagging indicators. The models fail to unambiguously recognize the strong link between performance management and organizational culture and climate.

Reniers et al. (44) presents a model to integrate a safety and security culture and climate with performance management, leading to company excellence achievement and leadership on safety as well as security. The suggested elements to this end are embedded in the Deming cycle. The proposed model – Improvement Diamond for Excellence Achievement and Leadership in Safety and Security (IDEAL S&S) – uses the P2T-model (explained in Section 3.3.2) to visualize the safety and security culture and climate in hazardous industries. ▶Fig. 3.28 illustrates the IDEAL S&S model.

In order to explain the IDEAL S&S model, a number of terms have to be defined. The IDEAL S&S model shows two fields of tension. The tension between optimal resources *versus* deployed recourses makes up the first field of tension. The term "resources" should be interpreted very broadly, e.g., money, knowledge, installations, know-how, people, etc. On one hand, optimal resources are those that are necessary to reduce a certain risk component until it is considered – some way or another – acceptable. They represent resources "as should be in ideal circumstances." On the other hand, deployed resources are/can be deployed in reality and thus in the real industrial setting of the plant – the situation "as is in real circumstances". In an ideal equilibrated situation, the optimal resources are equal to the deployed resources.

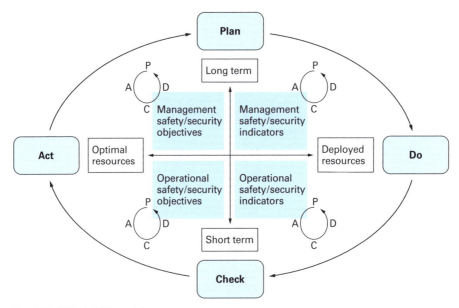

Fig. 3.28: IDEAL S&S model.

If the level of optimal resources lies above the level of the deployed resources, a potentially hazardous situation can occur. Conversely, if the deployed resources surpass the optimal ones, a company has wasted resources, because the risk level was already reduced below an acceptable level.

A second field of tension exists between short-term and long-term. Influencing a safety culture requires goals in the long-term that also need to be translated into manageable short-term goals. Hence, achieving the requirements for a company's safety and security culture is visualized in the (long-term) upper-part of the model whereas the safety and security climate (which is a snapshot of the company's culture) is represented by the (short-term) lower part of the model.

The IDEAL S&S model employs safety and security indicators. These indicators assign a qualitative or quantitative value to different safety and security climate and culture aspects. The long-term deployed resources are situated between the plan-phase and the do-phase of the Deming cycle and can be steered by using management safety and security indicators. The short-term deployed resources are situated between the do-phase and the check-phase and can be steered by using operational safety and security indicators. Both leading and lagging indicators should be used in both cases. Further information concerning developing leading and lagging (safety and security) indicators can be found in HSE (48) and Parmenter (49).

The IDEAL S&S model also uses safety and security objectives. The amounts of optimal resources have to make sure these objectives can be met. Management safety and security objectives and operational safety and security objectives, respectively, are quantitative figures or qualitative figures set to be achieved by company management or by business unit management for a specific management safety and security indicator, or a specific operational safety and security indicator, respectively. Management safety and security objectives (e.g., with a yearly, 2-yearly, or 5-yearly frequency) and operational safety and security objectives (e.g., with a weekly, monthly, or 3-monthly frequency) are used to control, remediate and continuously improve the organization's safety and security achievements, which ultimately lead to company excellence in safety and security.

Management indicators and objectives are used to influence and continuously optimize the company's safety and security culture (long-term approach), whereas operational indicators and objectives lead to constant organizational safety and security climate assessment and improvement (short-term approach). Indicators and objectives themselves should be continuously planned, implemented, checked, and adapted (if necessary) according to the Deming wheel of improvement.

Furthermore, all management and operational safety and security objectives and indicators should be worked out by using the three dimensions of the P2T-model, i.e. people, procedures and technology. Also, to integrate both safety and security into the model, the three dimensions are looked upon from both perspectives. Hence, people, procedures and technology are further categorized into different sub-dimensions (which are used for both safety and security) and various safety and/or security indicators are linked to each of these sub-dimensions. For all indicators, a set of minimum organization-specific safety and/or security objectives can be established by the user of the model. Based on the current literature [amongst others HSE (48) and OECD Guidance on Safety Performance Indicators (50)], a list of possible safety and security sub-dimensions may be identified. The sub-dimensions from ▶Tab. 3.2 and ▶Tab. 3.3 respectively, for the long-term and the short-term, are used for illustrative purposes.

Tab. 3.2: Safety and security long-term (strategic/management) sub-dimensions and indicator examples.

Long-term (strategic/management) sub-dimensions	Non-exhaustive list of indicator examples[1]
Technology	
Software, tools, etc. for safety and for security prevention, mitigation, emergency, etc. are used.	2-yearly budget available for safety software.
Technology (other than software) for safety and for security prevention, mitigation, emergency, etc. is used.	5-yearly budget available to maintain installations according to best available practices.
Technological knowledge and know-how of chemical processes, products, installations, etc. is regarded as essential.	2-yearly budget for training/educating personnel installation state-of-the-art knowledge.
Installations are safely and securely designed.	Percentage of installations that comply with international norms (DIN norms, ISO norms, etc.) within 2 years.
Procedures	
Existence of a company safety and security policy.	An external audit of the company's safety and security policy is carried out every 5 years.
Compliance with safety and security legislations at all times.	Number of legally prescribed safety procedures that are not fulfilled are <5% of all legally prescribed safety procedures; this is checked every 3 years.
Likelihood and severity of potential accidents are reduced to an acceptable level, using risk assessments and threat assessments.	Every 5 years the entire plant is checked by using Security Vulnerability Assessments (i.e., at least per 5 years, an SVA is carried out for every installation within the plant).
Safety and security improvements are driven by learning.	Number of incidents attributed to the same cause in 2 years.
A well-functioning safety management system is put in place, as well as a security management program.	A 3-yearly internal audit of the security management program is carried out.
Emergency preparedness and response procedures as well as Business Continuity Plans are in place.	The BCP is tested every 2 years.
Internal and external audits are drivers for change management and continuous improvement of safety and security within the company.	When an internal audit is performed, long-term recommendations for continuous improvement are required in the audit report.
Documentation of procedures.	A 4-yearly check is carried out by the security department whether all security procedures are written down, understandable, up-to-date and whether they can be easily consulted by its users.

(Continued)

(*Continued*)

Long-term (strategic/management) sub-dimensions	Non-exhaustive list of indicator examples[1]
People	
Involvement of top management in safety and security policy.	Overall 2-yearly budget assigned to security activities.
Involvement of employees in safety and security practices	Every 3 years, a security survey is organized among company personnel
Involvement with/of parties external to the company.	Every 2 years, contractor safety achievements are discussed with the contractors.
Employees are sufficiently competent concerning safety and security issues and they have adequate experience/expertise when needed.	A learning trajectory for employees exists within the company.
There is a very open sphere regarding safety; all employees are well-informed and are free to express ideas, discontentment, etc.; employees are involved in the decision-making process.	Score given to "corporate openness regarding safety" in a 2-yearly questionnaire.
Safety and security are the number one priorities and this is acknowledged by all employees.	Scores employees receive during safety observations using 360° feedback reviews carried out every 3 years.
There is mutual communication based on mutual trust concerning safety and security between all employees.	Score given to "mutual communication as regards security topics" in a 2-yearly questionnaire.
Employees are prepared for emergency situations.	Percentage of executed improvement propositions within 2 years resulting from emergency plan exercises.

[1] One indicator is given as an example per sub-dimension. Source: Reniers et al. (44).

Tab. 3.3: Safety and security short-term (operational) sub-dimensions and indicator examples (non-exhaustive).

Short-term (operational) sub-dimensions	Non-exhaustive list of indicator examples[1]
Technology	
Installations and chemical products and processes are regularly investigated for working as expected (safe and secure) and they are maintained wherever deemed needed on a regular basis.	Safety inspections are carried out at least every 6 months in every installation of the plant.
Risk assessment software and threat assessment software are employed for risk studies and they are regularly updated.	Access and gate control: the number of daily controlled persons.

(*Continued*)

(*Continued*)

Short-term (operational) sub-dimensions	Non-exhaustive list of indicator examples[1]
All technology (besides software) used within the company or business unit with regards to safety and security is regularly maintained and updated.	Ratio of corrective/predictive maintenance per month.
The workplace is safe and ergonomic to work in, and technological security measures (e.g., CCTV) are applied wherever necessary.	Weekly score given to workplace safety and housekeeping.
Procedures	
Company safety and security policy is clearly and unambiguously translated in operational safety and security goals per level and per business unit of the organization.	Number of yearly improvement proposals as a result of an internal audit in one of the installations of the company.
Hazardous substances are stored in a proper manner.	Daily housekeeping checklists are used for storing materials.
Working procedures, safety procedures, security requirements, installation specifications, etc. are well documented.	Percentage of standardization of security documentation, checked per 6 months.
Efficient and adequate (user-friendly) procedures have been developed for the staff to follow.	Percentage of procedures, still leading to difficulties and incidents, evaluated per year.
Company procedures and guidelines are in place regarding the frequency and the necessity/circumstances to use risk assessment or threat assessment software.	A frequency of SVAs to be carried out per installation is determined and the circumstantial conditions/approaches are described.
Procedures are in place to comply with existing safety and security regulations and to follow-up and comply with new safety and/or security legislation.	Degree to which existing security legislation is taken into account by company procedures is checked every 6 months.
The procedural expectations regarding safety and security are understood by all employees and everybody commits him- or herself to respect these procedural standards.	Degree to which working procedures of a business unit or installation are easy to understand and are followed is formally checked by the shift supervisor every 3 months (on top of daily informal checks).
All company safety and security procedures, guidelines, working instructions etc., are documented.	Level of standardization of security documents (procedures, guidelines, working instructions, etc.)
Company procedures and guidelines are in place regarding the frequency and the necessity/circumstances to have internal and/or external audits.	Number of scenarios (circumstances) for which a frequency of external audits is fixed.
The company has internal and external emergency plans as well as a business continuity plan.	Degree to which the external emergency plan is elaborated and tested for security situations (e.g., a terrorist attack).

(*Continued*)

(Continued)

Short-term (operational) sub-dimensions	Non-exhaustive list of indicator examples[1]
People	
Competence schemes and profiles are kept of all employees and a learning trajectory exists within the company.	Percentage of employees within an installation that has similar competences (and which leads to more flexibility within the work shift).
All employees are aware of all safety and security knowledge and know-how for carrying out his/her function.	Number of weekly visits of management to work-floor.
Every person fully understands his/her safety and security responsibilities and acts appropriately.	Number of monthly meetings where employees receive information and feedback about the importance of security.
The company can ensure a safe cooperation with – and operations of – contractors and outsiders.	Levels of satisfaction (questionnaire scores) regarding cooperation with external partners after yearly emergency exercises.
360° feedback reviews and open communication initiatives are taken within the company.	Daily operational staff meetings are held on safety.
Emergency exercises and drills are regularly carried out.	Every 3 months, a drill for security guards and dogs is held.

[1] One indicator is given as an example per sub-dimension. Source: Reniers et al. (44).

Every company should define its own safety and security sub-dimensions, which are eventually used to develop management safety and security indicators and objectives and operational safety and security indicators and objectives. The sub-dimensions presented in ▶Tab. 3.2 and 3.3 should be regarded as illustrative guidance.

3.8.1 Performance indicators

There should be clear and unambiguous objectives that can be used to evaluate the company policy. Conversely, only those items linked to the company's sub-dimensions, should be measured. Hence, an indicator should always be related to a sub-dimension. This way, it is also easier to: (i) change a sub-dimension and thus indicators linked to that sub-dimension, (ii) choose more or other indicator(s) for a sub-dimension, or (iii) change the objective of an indicator, etc.

Indicators should be "SMART":

- Specific and clearly defined.
- Measurable so that how the indicator is performing can be checked on a regular basis.
- Achievable so that each indicator provides a target that is stretching but not so extreme that it is no longer motivational (the indicator needs to have sufficient support).
- Relevant to the organization and what it is aiming to achieve.
- Time bound in terms of (realistic) deadlines or timing for when each indicator will be achieved.

Objectives can be formulated in different ways: as an absolute number (target numbers), as a percentage (decrease of x%, satisfy x% of criteria, satisfy x% of a checklist, etc.), or as a relative position to a benchmark (higher than the national mean, lower than the mean of the industrial sector, lower than one's own performances of the past x years, etc.).

The different parts of performance management as part of the organizational safety management system, are: dimension, sub-dimension, indicator, objective, result, deviation.

Mazri et al. (51) indicate that certain basic information, technical data, organizational information, and IT data are required for every indicator. ▶Tab. 3.4 provides an overview of the information required to ensure adequate use of performance indicators.

Moreover, as already mentioned, different types and levels of indicators exist. It is obvious that indicators should be defined for all three dimensions (and all their sub-dimensions). In Chapter 2, the risk sandglass was introduced, indicating the existence of positive as well as negative risks (the risks can, e.g., be determined by the combination of a SWOT analysis and more traditional risk analyses), and that positive risks should

Tab. 3.4: Performance indicators – information table.

General information	
Short name	Unique codified name of indicator.
Long name	Detailed name of indicator.
Description and purpose	What does and doesn't the indicator measure? (What would there possibly be confusion about?)
Source	Who issued this indicator?
References	Available reference document(s) concerning the indicator.
Nature	Qualitative, semi-quantitative, or quantitative.
Risk domains covered	Depending on the needs and the management systems implemented, a myriad of risk domains can be covered. For example, environment risks, health and safety risks, security risks, operational risks, process risks, occupational risks, quality risks, ethical risks, etc., or any combinations thereof. Note that a unique indicator may be more or less relevant for several domains.
Technical information	
Formula and unit	With what formula was the indicator value calculated (if applicable)?
Target value	Target value (to reach a predefined performance).
Minimal and maximal values	Describe the minimal and maximal limit values within which the indicator value may be considered as "acceptable". If the indicator value is out of these limit values, actions need to be taken.
Input data required	Information required to implement the formula described above (that led to the calculation of the indicator).
Frequency of measurement	What is the frequency with which this indicator should be measured (the periodicity of monitoring will influence on the level of resources required)?

(Continued)

3.8 The IDEAL S&S model

(Continued)

Technical information	
Related indicators	Indicators are part of a "network of indicators" monitoring different system components. The relationship(s) between the indicators should be mapped and a list of additional indicators providing extra information on the indicator under consideration should be drafted.
Organizational information	
Indicator reference person (or owner)	A reference person in the organization should be affected to each indicator. This person will be responsible for the quality of the whole process from data and information collection to interpretation and communication of the results.
Data provider(s) or registrator(s)	Person(s) need to be appointed to collect and deliver the required data/information (necessary input data).
Interpretation procedure	Person(s) need to be identified who are capable of, and who have the competence and the authority to, correctly interpreting the measured indicator value, and to translate this value into knowledge and insights.
Communication procedure	Person(s) within and outside the organization that should be informed about the indicator results, are to be identified. The method of communicating the results is to be determined.
Relevance assessment procedure	The relevance of any indicator should be questioned at regular time intervals and according to a predefined procedure.
IT-information	
Software availability	Existing software is listed that improves the use of the indicator or that makes it more easy.
Adequacy with existing/local information system	The configuration of existing software may facilitate the input of collected data, or it may complicate this process. This fact should be taken into account beforehand.

be maximized, while negative risks should be minimized. Thus, indicators should be elaborated for both positive and negative risks. Moreover, one should not be blind towards major catastrophes, and towards risks characterized with extreme high uncertainties. Hence, indicators should be identified for all three types of risks (types I–III). Furthermore, different decision levels require different indicators: management, process, and operational. Next to management and operational indicators, process indicators provide information on the working processes within the organization. Such indicators are very important, as they allow for gaining a system's view on the organization, which is indispensable for a safe organization (see also Section 3.10). Another distinction is the position of the indicator: measuring before an incident (proactive or leading indicator) or after an incident (reactive or lagging indicator). ▶Fig. 3.29 gives an overview of possible indicators to be elaborated. It should be obvious that indicators sometimes will be extremely hard to imagine and to think of, or will simply not exist.

Dimension 1	Positive risks	Type I	Management ind.	Leading indicators Lagging indicators
			Process ind.	Leading indicators Lagging indicators
			Operational ind.	Leading indicators Lagging indicators
		Type II & III	Management ind.	Leading indicators Lagging indicators
			Process ind.	Leading indicators Lagging indicators
			Operational ind.	Leading indicators Lagging indicators
	Negative risks	Type I	Management ind.	Leading indicators Lagging indicators
			Process ind.	Leading indicators Lagging indicators
			Operational ind.	Leading indicators Lagging indicators
		Type II & III	Management ind.	Leading indicators Lagging indicators
			Process ind.	Leading indicators Lagging indicators
			Operational ind.	Leading indicators Lagging indicators

Fig. 3.29: Types of indicators to be thought of and worked out.

The aim of performance indicators and performance management is obviously to improve the different domains of an organizational culture (people, procedures, and technology). Proactive and reactive monitoring allows taking measures in organizational processes, activities, attitudes, etc. ▶Fig. 3.30 visualizes the usefulness of performance management.

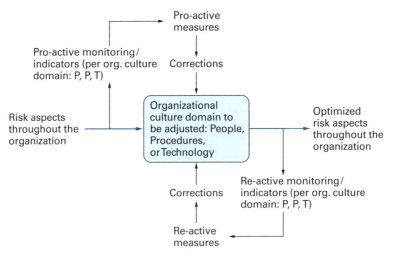

Fig. 3.30: Optimization of the risk aspects throughout the organization and performance management with reprint permission of Die Keure (52).

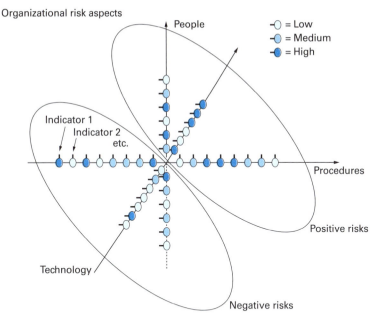

Fig. 3.31: Possible depiction for a quick visualization of people, procedures and technology with reprint permission of Die Keure (52).

For every organizational risk, a three-dimensional representation can be suggested, based on the three domains of organizational culture. ▶Fig. 3.31 shows a quick overview of the three individual organizational culture domains for negative as well as positive risks.

This way, a quick overview is provided of the most important/urging positive and negative risks, based on monitoring leading (for proactive monitoring) and lagging (for reactive monitoring) indicators. The levels at which actions are required, are indicated by light blue = no actions needed, medium blue = actions needed but not urgently (e.g., within 4 months), and dark blue = immediate action required.

> Performance management is a very powerful tool to systematically map the effectiveness with which every aim or goal (short-term or long-term) within the different safety and security dimensions and sub-dimensions of an organization is reached. It can also be used to prioritize actions, budget allocations, etc.

3.9 Continuous improvement of organizational culture

Based on social system theory, Wu et al. (53) studied the potential correlation between safety leadership, safety climate and safety performance. The results of the statistical analysis indicated that organizational leaders would do well to develop a strategy by which they improve the safety climates within their organizations, which will then have a positive effect on safety performance. We can therefore assume that employing the suggested IDEAL S&S model will lead to truly safer and more secure companies that will expose organizational safety and security leadership.

In summary, a safety culture is like a human relationship: it needs constant attention and constant labor to ensure its success in the short-term, but especially in the long-term. An (internal or external) audit only provides an idea of the safety climate at a certain point in time and does not give a true indication of the safety culture or of the "safety DNA" of an organization. To have a more accurate picture, safety performance management should be established, and safety audits should be repeated on a regular basis, and the audit results should be compared with each other and should be analyzed to make sure that there is continuous improvement over time. ▶Fig. 3.32 shows how this may be achieved.

> The way an organization deals with risks is also explained by the general attitude of the people working in the organization and its leadership. The way things are done regarding safety in an organization, is called "safety climate" (short-term) or "safety culture" (long-term). Different levels of maturity of safety culture can be discerned, from pathological over reactive, calculative, and proactive to generative. The difference between the proactive level and the generative level is mainly due to attitude and trust: in the generative level, people communicate freely about problems and trust each other within and across every layer of the organization. Things are done safely or not at all.

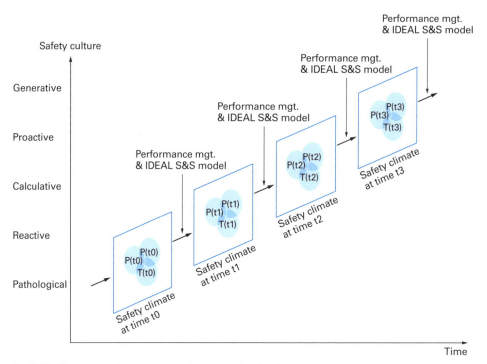

Fig. 3.32: Continuous improvement of organizational safety culture.

> Nonetheless, most best-of-class companies today are situated in the proactive level, where they actively look for potential problems, and adopt the attitude of "if you think safety is expensive, try an accident". However, they should really try achieving the generative level. The IDEAL S&S Model can be employed to get there in a realistic way.

3.10 High reliability organizations and systemic risks

3.10.1 Systems thinking

The different subsequent work process steps need to be understood, as well as the existing links between them. However, it is also very important to gain insights into the "logic behind the system." Such insights are essential to be able to interpret existing causalities at the level of organizational working processes. Some *general insights in systems thinking* are given hereafter. It should be noted that gaining insights within any organization requires "trial and error" procedures within the organization.

3.10.1.1 Reaction time or retardant effect

Every safety measure, taken based on risk management, has a certain "reaction time" – it takes a certain amount of time before the effect(s) of a measure become(s) apparent. It is important to know, or at least to have an idea of, this reaction time in order to avoid taking new measure(s) too quickly. Hence, a long-term vision, while taking safety and health measures and interpreting the results, needs to be supported by the insights of the working of the system, and of the long-term effects of measures on the system.

Senge (54) illustrates the idea by using the metaphor of a hot-water faucet. If one does not take into account the fact that 10 seconds are needed for the water to become hot, one might further turn open the hot water faucet, so that at the time the water becomes warm, it might be so hot that it hurts, leading the operator to turn the tap towards cold water; with the result of having cold water instead of hot water, and time is lost. Eventually, after a certain period of trial and error, the desired water temperature may be achieved, but a lot of time has been lost, dangerous situations may have occurred, and the work process is all but optimal.

3.10.1.2 Law of communicating vessels

The law of communicating vessels simply states that in physics matters are often linked to one another, and that in some way they have an impact on each other. This is no different for socio-economic systems such as organizations. Every measure or change within a system leads to changes and shifts within other parts of the system, and these domino-changes need to be identified and, if necessary, additional measures need to be taken.

The whole system at once, instead of different parts of the system, needs to be considered, and the relationships between the parts of the system need to be taken into account when making risk decisions. Hence, awareness of possible unexpected changes and continuous vigilance always need to be present.

3.10.1.3 Non-linear causalities

Linear causality is very hard to find in real life and in real industrial practice: almost never is an accident the result of one cause; on the contrary, an accident is usually (almost always) caused by the concurrence of circumstances and a variety of factors. One factor itself does often not suffice to cause an accident. Also, if another order of events had happened then the consequences of an accident might have been completely different. Moreover, cause and consequence regularly are not closely linked in space and time. Nonetheless, the urge to "think in linear causalities" by humans is very strong, and for risk managers it certainly has also been so in the past.

Instead of viewing reality as a static picture and, based on this picture, taking preventive measures, risk managers need to discern change patterns, looking at positive and negative feedback loops between events, and based on this improved perception of reality, take health and safety measures.

3.10.1.4 Long-term vision

Research indicates that business failures more often originate from bad adaptation of a business to slowly emerging threats, instead of being caused by sudden threats (54). Insidious, latent (long-term) problems usually do not receive the attention they deserve; while on the contrary, short-term failures leading to problems often do.

Because analyses are limited in space and time, gradually emerging failures and problems are much more difficult to detect. Increasing space and time while analyzing a certain part of reality improves the perception of this reality piece. An improved perception of reality leads to better decisions.

3.10.1.5 Systems thinking conclusions

All previously mentioned "laws of systems thinking" indicate the need for insights in a system, thereby considering all relevant system parts, their relationships, and their interdependences. Often, "easy solutions" are sought, looking for symptoms instead of underlying causes and structures. This way, the "visible problem" is solved for a short time, cannot be seen anymore, and all seems to be well. Of course, in reality, the problem is still present, and it will reappear later in time, often with more persistence and possibly somewhere else in a system. Also, the problem has often become that of someone else, who finds it harder to solve the problem because of the superficial actions taken previously.

Such a non-systems thinking behavior is strongly present in people, and knowledge and reasoning is therefore necessary to implement true systems thinking behavior in organizations. This can be illustrated by a fireman who is considered a hero when distinguishing fire at risk of his own life, but who is considered a stickler when implementing fire safety regulations (55).

3.10.2 Normal accident theory (NAT) and high reliability theory (HRT)

Two schools of thought exist on how to prevent major accidents: the high reliability theory (HRT), and the normal accident theory (NAT). HRT believes that with intelligent organizational design and management techniques a company can compensate for weaknesses within the organization and guarantee accident-free operations. NAT, on the contrary, believes that major accidents are inevitable, and suggests that complex organizations make decisions in ways that are different from the rational models used by the HRT theorists.

According to the HRT, four aspects lead to zero-accident safety: leadership that prioritizes safety objectives as an organizational goal, high levels of human and non-human redundancy, an organizational setting of high reliability with decentralized authority and a line level culture of reliability and training, and an approach to trial and error learning where organizations learn from (internal and external) experiences.

According to NAT theorists however, complex organizations may work hard to maintain safety and reliability but, despite all efforts, major accidents will be a "normal" result or an integral characteristic of the system. NAT theorists provide some causes for this: different individuals at different levels of an organization may hold conflicting goals, there may be important communication problems, the processes may not be fully understood by the employees, essential risk knowledge may have left the company with retirement, etc.

In any case, the levels of vulnerability and resiliency of organizations are very important in order to decide on how to deal with prevention, mitigation, emergency management, etc. within an organization. Furthermore, the type of activity of an organization has an impact on its levels of resiliency and vulnerability. Perrow (56) discerns two important dimensions in organizations: "interactions" and "coupling".

Interactions can be linear or complex. Linear interactions may be complicated, but they are always clear and visible to some extent. Complex interactions on the other hand can be unexpected or even incomprehensible. Ideally, systems should be made as linear as possible. However, this is not always possible in the (complex) industrial world of today. Systems subject to complex interactions are not always more dangerous or hazardous, but they are less predictable.

Some organizations or systems (within organizations) are tightly organized according to fixed procedures and with strong interdependence. Perrow (56) calls this "tightly coupled systems". In this type of organization, taking care of errors, disturbances, and failures are built in, without many possible ways to deviate from suggested scenarios and solutions. In other "loosely coupled" organizations, there is much more possibility to improvise and to use alternative solutions.

NAT theory, elaborated by Perrow (56,57) explains "system accidents" by using both these dimensions and their implications (such as non-linearities, reinforcing causalities, complexities and interactions, and strong structural links ("tight coupling"). Organizations can be divided into four quadrants, as illustrated in ▶Fig. 3.33. The nuclear industry and to a lesser extent the chemical industry are examples of organizations belonging to complex tightly coupled working environments. Nonetheless, insights gained from the NAT can also be applicable for non-complex working environments.

The theory boils down to people (regardless of the environment in which they work) understanding and appreciating the possibility of minor faults and failures interacting in unexpected ways, and, through the existence of structural relationships between system components, leading to a cascade of faults/failures, eventually leading to an accident. The higher the complexity of a system, the more it depends on coincidence and unexpected events, from both a positive and a negative point

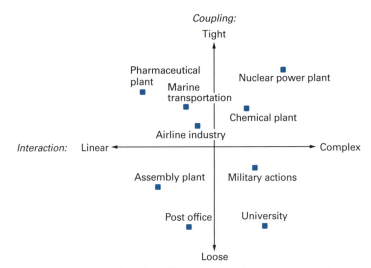

Fig. 3.33: Interaction versus coupling for different industrial sectors [inspired by (58)].

of view. A failure in a linear system (e.g., an assembly line) can usually be anticipated, and it is understandable and visible. Such a failure can be avoided by taking the correct preventive measures, being correctly executed and maintained. Complex interactive systems can be subject to different failures, of which each of the failures separately may not cause any problems because the correct preventive measures are in place. However, it may be the case that because of unexpected and incomprehensible interactions between different simultaneous failures of the complex system, the separate preventive measures are bypassed or nullified. If the system, next to being complex, would be tightly linked as well, the failures may get out of control, and an avalanche of failures may arise, resulting in a major system failure and a subsequent major accident.

In summary, HRT theorists maintain that organizations can be designed and managed to near-perfect reliability and safety records. NAT theorists maintain that, in these same organizations, major "system" accidents are inevitable because of "system faults". The difference in perception between the theorists of HRT and NAT boils down to how much influence each one believes that risk management has on the performance of the organization.

In this regard, the difficulty of having a technological disaster should be stressed. For example, chemical plants with catastrophic potential have been around for more than a century, yet there have been few disasters that claimed thousands of lives. There has, e.g., only been one Bhopal. Nonetheless, the US EPA estimated in 1989 that in the previous 25 years there were 17 releases of toxic chemicals in volumes and levels of toxicity exceeding those that killed many thousands of lives in Bhopal. Mostly because of sheer luck only five people were killed in these accidents (59). As Perrow (60) indicates, the reason is the flip side of NAT: just as it takes the right combination of failures to defeat all the safety devices, so does it take the right combination of circumstances to produce a true disaster.

3.10.3 High reliability organization (HRO) principles

Organizations capable of gaining and sustaining high reliability levels are called "high reliability organizations" or HROs. Despite the fact that HROs operate hazardous activities within a high-risk environment, they succeed in achieving excellent health and safety figures. Hence, they identify and correct risks very efficiently and effectively.

A typical characteristic of HROs is collective mindfulness. Hopkins (61) also indicates that HROs organize themselves in such a way that they are better able to notice the unexpected in the making and halt its development. Hence, collective mindfulness in HROs implies a certain approach of organizing themselves. Five key principles are used by HROs to achieve such mindful and reliable organization [see also (62)].

The first three principles mainly relate to anticipation, or the ability with which organizations can cope with unexpected events. Anticipation concerns disruptions, simplifications, and execution, and requires means of detecting small clues and indications, with the potential to result in large, disruptive events. Of course, such organizations should also be able to decrease, to diminish or to stop the consequences of (a chain of) unwanted events. Anticipation implies the ability to imagine new, non-controllable situations, which are based on little differences with well-known and controllable situations. HROs take this into account in principles 1, 2, and 3.

Whereas the first three principles relate to pro-action, the fourth and fifth focus on reaction. It is evident that if unexpected events happen despite all precautions taken, the consequences of these events need to be mitigated. HROs take this into account in principles 4 and 5.

3.10.3.1 HRO principle 1: targeted at disturbances

This principle asserts that HROs are very actively, and in a proactive manner, looking for failures, disturbances, deviations, inconsistencies, etc. because they realize that these phenomena can escalate into larger problems and system failures. They achieve this goal by urging all employees to report (without a blame culture) mistakes, errors, failures, near-misses, etc. HROs are also very much aware that a long period of time without any incidents or accidents may lead to employee complacency, and may thus further lead to less risk awareness and less collective mindfulness, eventually leading to accidents. Hence, HROs rigorously ensure that such complacency is avoided at all times.

3.10.3.2 HRO principle 2: reluctant for simplification

When people – or organizations – receive information or data, there is a natural tendency to simplify or to reduce it. Parts of the information considered as non-important or irrelevant, are omitted. Evidently, information that may be perceived as irrelevant might in fact be very relevant in order to avoid incidents or accidents. HROs will therefore question the knowledge they possess from different perspectives and at all times. This way, the organizations try to discover "blind spots" or phenomena that are hard to perceive. To this end, extra personnel (as a type of human redundancy) are used to gather information.

3.10.3.3 HRO principle 3: sensitive towards implementation

HROs strive for continuous attention towards real-time information. All employees (from front-line workers to top management) should be very well-informed about all organizational processes, and not only about the process or task they are responsible for. They should also be informed about the way that organizational processes may fail and how to control or repair such failure.

To this end, an organizational culture of trust between and among all employees is an absolute must. A working environment in which employees are afraid to provide certain information, e.g., to report incidents, will result in an organization being lacking in information, and in which efficient working is impossible. An "engineering culture", in which quantitative data/information is much more appreciated than qualitative knowledge/information, should also be avoided. HROs do not distinct between qualitative and quantitative information.

HROs are also sensitive towards routines and routine-wise handling. Routines can be dangerous when leading to mindlessness and distraction. By installing job rotation and/or task rotation in an intelligent way, HROs try to prevent such routine-wise handling.

Furthermore, HROs view near-misses and incidents as opportunities to learn. The failures that go hand-in-hand with the near-misses always reveal potential (otherwise hidden) hazards, hence such failures serve as an opportunity to avoid future similarly caused incidents.

3.10.3.4 HRO principle 4: devoted to resiliency

HROs define "resiliency" as "the capacity of a system to retain its function and structure, regardless of internal and external changes". The system's flexibility allows it to keep on functioning, even when certain system parts do not function as required anymore. An approach to ensure this, is that employees organize themselves into ad hoc networks when unexpected events happen. These can be regarded as temporary informal networks capable of supplying the required expertise to solve the problems. When the problems have disappeared or are solved, the network ceases to exist.

3.10.3.5 HRO principle 5: respectful for expertise

Most organizations are characterized by a hierarchical structure with a hierarchical power structure, at least to some degree. This is also the case for HROs. However, in HROs, the power structure is no longer valid in unexpected situations in which certain expertise is required. The decision process and the power are transferred from those at the top of the hierarchy (in normal situations) towards those with the most expertise regarding certain topics (in exceptional situations).

> Two theories exist of accidents in large-scale systems with catastrophic potential: HRT and NAT. HRT believes that organizations can learn from operating and regulatory mistakes, put safety first and empower lower levels, thereby making risky items quite safe. NAT suggests that, no matter how hard organizations try, there will be serious accidents because of the interactive complexity and tight coupling of most risky systems.

3.10.4 Risk and reliability

The safety of an organization, especially a complex tightly coupled organization, depends on a variety of factors, evidently starting with good design of processes, equipment, installations, etc. As a process, installation or equipment, no matter how well designed, cannot operate indefinitely without intervention, the degree of safety depends on the maintenance procedures and on actions intended to keep the organization safe. Reliability engineering, as a part of risk engineering, is concerned with analyzing the *reliability, availability and maintainability* (RAM) of a safety system. Several issues related to performance must be considered: hardware and software failures, human errors, incorrect operating procedures, and also the interactions between these. But what are the differences between those terms? Definitions are given by Sutton (63):

- The reliability of a component or of a system is the probability that it will perform a required function without failure under stated conditions for a stated period of time.
- The availability of a repairable system is the fraction of time that it is able to perform a required function under stated conditions.
- The maintainability of a failed component or system is the probability that it is returned to its operable condition in a stated period of time under stated conditions and using prescribed procedures and resources.

The difference between reliability and availability arises because reliability does not account for the possibility that a given system can be repaired after its failure. This indicates that the reliability in function of a time t, noted as $R(t)$, predicts the time t until the system has undergone its first failure (thus reliability refers to the *first* system failure), whereas the system may have failed in the past but has been repaired so that it is operational at time t with predicted availability $A(t)$. Reliability can also be seen as the complement of the failure probability $F(t)$, hence $R(t) = 1 - F(t)$.

A system with redundant subsystems can exhibit sub-system failures without system failure. For an availability analysis, the on-going repair actions continue. Availability is thus used for systems and reliability is employed for components of those systems.

Maintainability is the ability of a system component to be restored to a state in which it can perform its intended function when the maintenance is performed under prescribed procedures. It involves actions typically performed according to procedures established by the manufacturer of the component.

For further information on reliability engineering, including probability theory, Boolean algebra, failure rates probability distributions, mean time between failures (MTBF), mean time to failures (MTTF), mean time to repair (MTTR), mean downtime (MDT), etc., we refer to specialized literature such as Sutton (63), Smith (64), Lee and McCormick (64), Zio (65), and others.

One important and much-quoted concept from reliability theory, is the "bathtub distribution". The bathtub curve seeks to describe the variation of failure rate of components during their life. ▶Fig. 3.34 illustrates this generalized relationship.

From a conceptual point of view, the bathtub curve is interesting as it clearly demonstrates the usefulness of RAM for safety purposes. The failures shown in the first part of the curve, where the failure rate is decreasing, are called "early failures"

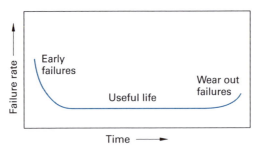

Fig. 3.34: Bathtub curve.

or "infant mortality failures". They are usually related to manufacture and quality assessment, e.g., connections, joints, dirt, impurities, cracks, insulation or coating flaws, incorrect adjustment, incorrect positioning, etc. The middle portion is referred to as the "useful life" and it is assumed that failures exhibit a constant failure rate, i.e. they occur at random. The failures from this region are usually assumed to be stress-related. The latter part of the curve describes the wear out failures and it is assumed that failure rate increases as the wear out mechanisms accelerate. Such failures result from corrosion, oxidation, breakdown of insulation, atomic migration, friction wear, fatigue, etc.

3.11 Accident reporting

This section deals with the question how and in what ways incident and accident figures can be used as reactive (lagging) indicators to measure safety effectiveness within an organization. It is an important topic, because many organizations still use this type of performance indication to have an idea of how well – or how bad – they are performing regarding safety. Several types of metrics are used. If we discern, as the Bird pyramid (▶Fig. 3.5) suggests, serious accidents from minor injuries, different metrics for different kind of accidents are used.

For serious accidents, which we may define as occurrences that resulted in a fatality, permanent disability or time lost from work of one day/shift or more, usually the lost-time injury frequency rate (LTIFR) is used. The LTIFR is the number of lost-time injuries per million hours worked, calculated using this equation:

LTIFR = Number of lost-time injuries over the accounting period × 1,000,000 ÷ Total number of hours worked in accounting period

Hence, the LTIFR is how many lost-time injuries (LTI) occurred over a specified period per 1,000,000 (or some other number; 100,000 is also often used) hours worked in that period. Mostly, the accounting period is chosen to be 1 year. By counting the number of hours worked, rather than the number of employees e.g., discrepancies that may be caused in the incidence rate calculation by part-time workers and overtime are avoided. However, this metric using the employees instead of the hours worked – the lost-time injury incidence rate (LTIIR) – is also used in many organizations. To calculate the LTIIR, which is the number of LTIs per 100 (or whatever figure you want) employees, the following equation are used:

LTIIR = Number of lost-time injuries over the accounting period × 100 ÷ Average number of employees in accounting period

Next, the severity rate, which takes into account the severity per accident. Depending on how this is expressed you will need at least the information from above and the number of work days lost over the year. Often the severity rate is expressed as an average by simply dividing the number of days lost by the number of LTIs. Another way of calculating the severity rate, the lost-time injury severity rate (LTISR), is using the following equation (the figure 1,000,000 may be replaced by any other figure; it just tells us that the LTISR in this case is expressed per million hours worked):

LTISR = Number of work days lost over the accounting period × 1,000,000 ÷ Total number of hours worked in accounting period

Also, the medical treatment injury (MTI) frequency rate is often measured. This frequency rate measures how often medical treatment injuries are occurring. It is expressed as the number of medical treatment injuries per million hours worked:

MTIFR = Number of medical treatment injuries over the accounting period × 1,000,000 ÷ Total number of hours worked in accounting period

Finally, the total recordable frequency injury rate (TRIFR) measures the frequency of recordable injuries, i.e., the total number of fatalities, lost-time injuries, medical treatment injuries and restricted work injuries occurring per million hours worked:

TRIFR = Number of recordable injuries (fatalities + lost-time injuries + medical treatment injuries + restricted work injuries) over the accounting period × 1,000,000 ÷ Total number of hours worked in accounting period

For more information regarding accident reporting, see HSE (67).

3.12 Conclusions

On the one hand, risk management principles cannot be summarized easily or in simple terms: the various ways to handle and manage risks are diverse and focused on the type and the characteristics of the risks at hand. On the other hand, although risk is an abstract concept, our natural human understanding of "risk" and especially of decision-making related to risk, is pretty sophisticated. As an example, a predictable or expected loss is not the same as an unexpected loss or a catastrophic loss: we all intuitively understand this. We also understand that different risks require different approaches and different actions. We even have our own ideas about what approach would be suitable for which circumstances. But these ideas are not always right, and in fact a lot of our intuitive thoughts about risks cannot be trusted. Dealing with risks in our era is something that should be learned, and that requires experience and expertise. In this chapter, different ways and models to deal with different types of risks are presented. Although rather generic and theoretical, the models are easy to understand, and they can be applied on an operational level quite easily. Most importantly, they provide profound and valuable insights into the abstract world of risks, incidents, and accidents, and the abstract and concrete requirements to deal with them.

References

1. Bernstein, P.L. (1998) Against the Gods. The Remarkable Story of Risk. New York: John Wiley and Sons.
2. Vaughan, E.J. (1997) Risk Management. New York: John Wiley and Sons.
3. Fuller, C.W., Vassie, L.H. (2004) Health and Safety Management. Principles and Best Practice. Essex, UK: Prentice Hall.
4. ISO 31000:2009 Risk Management Standard – Principles and Guidelines, International Organization for Standardization, 2009.
5. Heinrich, H.W. (1950) Industrial Accident Prevention. 3rd edn. New York: McGrawHill Book Company.
6. Bird, E., Germain, G.L. (1985) Practical Loss Control Leadership, The Conservation of People, Property, Process and Profits. Loganville, GA: Institute Publishing.
7. James, B., Fullman, P. (1994) Construction Safety, Security and Loss Prevention. New York: Wiley Interscience.
8. Hopkins, A. (2010) Failure to Learn. The BP Texas City Refinery Disaster. Sydney, Australia: CCH Australia Ltd.
9. OGP (2005) Human Factors. London: International Association of Oil and Gas Producers.
10. Reniers, G., Dullaert, W. (2007) Gaining and Sustaining Site-integrated Safety and Security in Chemical Clusters. Zelzate, Belgium: Nautilus Academic Books.
11. Reason, J.T. (1997) Managing the Risks of Organizational Accidents. Aldershot, UK: Ashgate Publishing Limited.
12. Heinrich, H.W. (1950) Industrial Accident Prevention, 3rd edn. New York: McGrawHill Book Company.
13. Heinrich, H.W. (1959) Industrial Accident Prevention, 4th edn. New York: McGrawHill Book Company.
14. Heinrich, H.W., Petersen, D., Roos, N. (1980), Industrial Accident Prevention, 5th edn. New York: Mc GrawHill Book Company.
15. Powell, R.L. (1996) Process Safety and Control Systems Integrity. International Conference and Workshop on Process Safety Management and Inherently Safer Processes, October 8-11, 1996, Orlando, Florida. pp. 227–241. New York: American Institute for Chemical Engineers.
16. Reniers, G.L.L. (2010) Multi-plant Safety and Security Management in the Chemical and Process Industries. Weinheim, Germany: Wiley-VCH.
17. CCPS, Center for Chemical Process Safety (1996) Inherently Safer Chemical Processes. A Life Cycle Approach. New York: American Institute of Chemical Engineers.
18. International Standard IEC-61511 (2003) 1st edn.
19. Gardner, R.J., Reyne, M.R. (1994) Selection of Safety Interlock Integrity Levels. Wilmington, DE: Dupont Engineering.
20. CCPS, Center for Chemical Process Safety (2003) Guidelines for Analyzing and Managing the Security Vulnerabilities of Fixed Chemical Sites. New York: American Institute of Chemical Engineers.
21. Vincoli, J.W. (1994) Basic guide to Accident Investigation and Loss Control. New York: John Wiley and Sons.

22. Ale, B.J.M. (2009) Risk: An Introduction. The Concepts of Risk, Danger and Chance. Abingdon, UK: Routledge.
23. CCPS, Center for Chemical Process Safety (2000) Guidelines for Chemical Process Quantitative Risk Analysis, 2nd edn. New York: American Institute of Chemical Engineers.
24. Ale, B.J.M. (2005) Living with risk: a management question. Reliab. Eng. Syst: Safe 90:196–205.
25. Reniers, G., Ale, B.J.M., Dullaert, W., Foubert, B. (2006) Decision support systems for major accident prevention in the chemical process industry: a developers' survey. J. Loss Prev. Process Ind. 19:604–662.
26. OHSAS 18001:2007 (2007) Occupational Health and Safety Management Systems – Requirements. London, UK: British Standardisation Institute.
27. Le Ray, J. (2010) Gérer les risques: Pourquoi? Comment? Saint-Denis-La Plaine, France: Afnor.
28. Schein, E.H. (2010) Organizational Culture and Leadership. San Francisco, CA: Jossey-Bass.
29. Irani, Z., Beskese, A., Love, P. (2004) Total quality management and corporate culture: constructs of organizational excellence. Technovation 24:643–650.
30. Guldenmund, F.W. (2007) The use of questionnaires in safety culture - an evaluation. Safety Sci. 45(6):723–743.
31. Guldenmund, F.W. (2010) Understanding and Exploring Safety Culture. Oisterwijk, The Netherlands: Boxpress.
32. Zohar, D. (1980) Safety climate in industrial organizations: Theoretical and applied implications. J. Appl. Psychol. 65:96–102.
33. Niskanen, T. (1994) Safety climate in the road administration. Safety Sci. 17:237–255.
34. Coyle, I.R., Sleeman, S.D., Adams, N. (1995) Safety Climate. J. Safety Res. 26:247–254.
35. Diaz, R.I., Cabrera, D.D. (1997) Safety climate and attitude as evaluation measures of organizacional safety. Accident Anal. Prev. 29:643–650.
36. Seo, D.-C., Torabi, M.R., Blair, E.H., Ellis, N.T. (2004) A cross-validation of safety climate scale using confirmatory factor analytic approach. J. Safety Res. 35:427–445.
37. Guldenmund, F.W. (2000) The nature of safety culture: a review of theory and research. Safety Sci. 34:215–257.
38. Van Leeuwen, M. (2006) De Veiligheids barometer, dissertation, University of Twente.
39. Reason, J. (1998) Achieving a safe culture: theory and practice. Work Stress 12:293–306.
40. Cooper, M. (2000) Towards a model of safety culture. Safety Sci. 36:111–136.
41. Hale, A.R. (2000) Culture's confusions. Safety Sci. 34:1–14.
42. Wiegmann, D.A., Zhang, H., von Thaden, T.L., Sharma, G., Mitchell, A.A. (2002) A Synthesis of Safety Culture and Safety Climate Research. Technical Report ARL-02-3/FAA-02-2 for the Federal Aviation Administration, Atlantic City International Airport, NY.
43. Mohamed, S. (2003) Scorecard approach to benchmarking organizational safety culture in construction. J. Construct. Eng. Manag.129:80–88.
44. Reniers, G., Cremer, K., Buytaert, J. (2011) Continuously and simultaneously optimizing an organization's safety and security culture and climate: the improvement diamond for excellence achievement and leadership in safety and security (IDEAL S&S). J. Clean. Prod. 19:1239–1249.
45. CCPS, Center for Chemical Process Safety (2007) Human Factors Methods for Improving Performance in the Process Industries. Hoboken, NJ: American Institute of Chemical Engineers.
46. Kletz, T.A. (2001) An Engineer's View of Human Error. 3rd edn. Rugby, UK: Institute of Chemical Engineers.
47. Thompson, A.A., Strickland, A., Gamble, J.E. (2007) Crafting and Executing Strategy – Text and Readings. New York: McGrawHill/Irwin.
48. HSE, Health and Safety Executive (2006) Developing Process Safety Indicators. A Step-by-step Guide for Chemical and Major Hazard Industries. Sudbury, UK: HSE Books.
49. Parmenter, D. (2007) Key Performance Indicators. Developing, Implementing, and using winning KPIs. Hoboken, NJ: John Wiley and Sons.

50. OECD (2003) Guidance on Safety Performance Indicators. Paris, France: OECD Environment, Health and Safety Publications.
51. Mazri, C., Jovanovic, A., Balos, D. (2012) Descriptive model of indicators for environment, health and safety management. Chem. Eng. Trans. 26:465–470.
52. Van Heuverswyn, K., Reniers, G. (2012) Performance Management 4: Van safetymanagement naar performant welzijnsmanagement. Brugge, Belgium: Die Keure.
53. Wu, T.C., Chen, C.H., Li, C.C. A correlation among safety leadership, safety climate and safety performance. J. Loss Prevent. Proc. 21:307–318.
54. Senge, P. (1992) De Vijfde Discipline, De kunst en de praktijk van de lerende organizatie. Schiedam, The Netherlands: Scriptum Management.
55. Bryan, B., Goodman, M., Schaveling, J. (2006) Systeemdenken, Ontdekken van onze organizatiepatronen. Den Haag, The Netherlands: Sdu Uitgevers.
56. Perrow, C. (1984) Normal Accidents. Living with High Risk Systems. New York: Basic Books.
57. Perrow, C. (1999) Normal Accidents. Living with high-tech risk technologies. Princeton, NJ: Princeton University Press.
58. Zanders, A. (2008) Crisis management. Organizaties bij crises en calamiteiten. Bussum, The Netherlands: Uitgeverij Coutinho.
59. Shabecoff, P. (1989) "Bhopal Disaster Rivals 17 in US". New York Times, 30 April.
60. Perrow, C. (2006) The limits of safety: the enhancement of a theory of accidents. In: Key Readings in Crisis Management. Systems and Structures for Prevention and Recovery, Smith & Elliott, editors. Abingdon, UK: Routledge.
61. Hopkins, A. (2005) Safety, Culture and Risk. The Organizational Causes of Disasters. Sydney, Australia: CCH Australia Ltd.
62. Weick, K.E., Sutcliffe, K.M. (2007) Managing the Unexpected. Resilient Performance in an Age of Uncertainty. 2nd edn. San Francisco, CA: John Wiley and Sons.
63. Sutton, I. (2010) Process Risk and Reliability Management. Operational Integrity Management. Oxford, UK: Elsevier.
64. Smith, D.J. (2011) Reliability, Maintainability and Risk. Practical methods for engineers, Oxford, UK: Butterworth-Heinemann.
65. Lee, J.C., McCormick, N.J. (2011) Risk and Safety Analysis of Nuclear Systems. Hoboken, NJ: John Wiley and Sons.
66. Zio, E. (2012) An Introduction to the Basics of Reliability and Risk Analysis. Singapore: World Scientific Publishing.
67. Health and Safety Executive (HSE) (2009) HSE Event Injury Illness Classification Guide. Sudbury, UK: HSE Books.

4 Risk diagnostic and analysis

4.1 Introduction to risk assessment techniques

A risk assessment is an important step in protecting workers and business, as well as for complying with the law. It helps to focus on the risks that really matter at the workplace – the ones with the potential to cause real harm. Currently, over 100 risk analysis techniques are available in the literature. Most of them identify initiating events (causes), consequences, safeguards, and recommendations. The main difference between the methods is in the way they approach the identification of causes or consequences. Empirical research revealed that the four most constituting techniques are hazard and operability studies (HAZOP), failure mode and effects analysis (FMEA) or failure mode, effects and criticality analysis (FMECA), what-if analysis and the risk matrix (1). Other techniques that are actually mainly used in the process industries are event trees analysis, fault trees analysis, human reliability analysis and check lists. Some of them will be discussed here, but for more information we refer to Groso et al. (2). Preliminary hazard analysis (PHA) is also presented as a relatively simple and inexpensive technique that still provides meaningful results.

Through the use of systematic methods, risk analysis aims to objectify the risks incurred by a system (globally speaking). In this sense, it only serves as a starting point and as a support to the decision-taking process relative to the acceptable level of risk, which is risk management. The latter is based on a wider range of criteria consisting of subjective factors. As mentioned in the previous chapter, risk management is the political (decision-making) process to deal with risks, while risk analysis and risk assessment are only technical approaches to have a good notion of risks. Hence:

Risk management = risk analysis + risk assessment + risk communication + business continuity planning + preventive training + incident analysis + influencing risk perception + …

The context of a particular risk analysis can be defined by two main situations:

- Cases involving a technical object, e.g., a plane, vehicle or machine.
- Cases involving more complex systems, e.g., industrial plants, agricultural and urban installations. These systems also include machines and other technical objects but in this case they will be closely linked with their environment.
- The methods and tools for risk analysis differ for the two situations:
- In the first case, one mainly uses classical tools for safety engineering (preliminary risk analysis and fault or event trees analysis).
- In the second case, these tools only allow an individual analysis of the system compartments, mainly technical objects, and it would be necessary to refer to complete methods that include these tools but are also capable of organizing their implementation.

The available tools can be classified into two categories:

- Semi-empirical tools such as PHA, which eventually lead to the development of grids issued from experiment results, FMEA and FMECA, HAZOP and functional analysis.
- Logical tools such as tree charts (fault tree analysis and cause-consequence charts) and logical models such as Markov chains (relations between probabilities and partial differential equations) or Petri networks.

All these tools allow approaches through mathematical calculations. The implementation of these tools can present certain difficulties, as most of them originate from reliability analysis of objects or object elements and thus are not fully adequate for complex risk analysis. Furthermore their implementation requires information that does not come from the tools themselves.

4.1.1 Inductive and deductive approaches

Two main approaches to risk analysis exist, inductive and deductive, also called, respectively, bottom-up and top-down.

- The deductive methodologies analyze the causes of an adverse event (accident) by answering the question, "How is it that this event may occur (search for causes)?"
- The inductive methodologies analyze the consequences of failure (initiating event) and answer the question "What adverse events can result in (search for consequences)?"

Inductive analyses are headed in the direction of accidental process, while the deductive analyses back up to this process (see ▶Fig. 4.1).

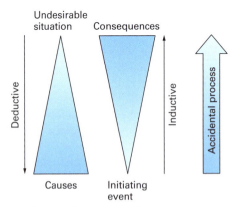

Fig. 4.1: Inductive and deductive methods.

There is also a complementary method that classifies risk analysis methodologies depending on the purpose fields to the search:

- The prospective analyses use a preventive approach. They can improve the system in its design phase and thus form part of the philosophy "prevention rather than cure".
- Conversely, retrospective studies investigate accident scenarios that have occurred when trying to find the causes and improving the system afterwards.

Prospective analyses rather rely on inductive methods, while the retrospective analyses almost always use deductive approaches.

4.1.2 General methods for risk analysis

In this chapter, selected methods of risk analysis will be briefly presented trying to cover the field of risk analysis. None of them are fully adequate for every situation but one need to use the method best suited to each study. The techniques can be classified in three distinct groups: basic methods, static methods and dynamic methods.

Among the well-known basic methods there is:

- Functional analysis
- Preliminary risk analysis
- Failure mode, effects and criticality analysis
- Hazard and operability study

These basic methods are often used during preliminary stages and proceeding towards a first global and high-level analysis. Depending on the complexity of the system, a more detailed analysis may or may not be required. Most basic methods are inductive (cause $\to$ effect) and do not require much more than a pencil and a table chart: there are no underlying mathematical models involved. These methods are indispensable for understanding how the system works and for correctly identifying the risks regardless of the type of system and the nature of the study.

The most common static methods are:

- Reliability block diagram (RBD)
- Fault tree analysis
- Event tree analysis

These methods, compared to the previous basic ones, allow an analysis from a structural (topological) point of view. This is obtained through Boolean mathematical models, which are static as they cannot model temporal effects on the system. To simplify, we can say that this approach only allows a logical representation of the system at a certain time and does not take account of changes in the system with time. Among the cited methods, fault tree analysis should be mentioned in particular because it is the only deductive (effect $\to$ cause) tool in reliability engineering.

The most well-known dynamic methods are:

- Markov chains.
- Stochastic Petri nets
- Bayesian belief networks.

These methods, with which one must now include formal languages such as AltaRica, were developed to take into account the temporal and compartmental effects that cannot be appropriately covered by static models. While the Markov approach is analytical, the stochastic Petri net method requires calculations through Monte Carlo simulation. Bayesian belief networks will be developed later in this chapter.

A question that is immediately raised by the reader is probably: "Why so many methods?" (▶Tab. 4.1 provides a non-exhaustive overview of risk analysis methods for engineering projects.)

Among the several existing methods, it is critical to select the one best suited to the study being carried out. An inappropriate selection can result in incorrect results or significant loss of time.

The choice of method mainly depends on:

- The nature and requirements of the study.
- The amount of knowledge of the system.
- Availability of quantitative data.
- Availability of time and resources.

The key questions to be asked before choosing a method should be:

- What are the aims and what is the scope of the analysis?
- Is the analysis prospective or retrospective?
- Is the analysis specific (linked to an event, failure, etc.) or does it consider the system as a whole?
- What is the required depth of analysis?
- How much time and resources are available?
- How well is the system known and which data are available?

> There are no universal methods, the most suitable method should be selected for each situation, and neither is there a specific guideline for selecting a method, rather the situation should be globally studied in order to determine the appropriate method.

Furthermore, certain methods are more adequate than others depending on the phase of the project. Aims as well as the required knowledge and data vary according to the phase. The following table illustrates the possible choice of suitable methods in engineering projects depending on the project phase, and specifies the aims, at what point in the process the risk analysis should be used, and which documents are required as a support.

▶Tab. 4.2 summarizes the main characteristics of different risk analysis techniques. Not all of them will be developed in this chapter, but this will provide some information about pros and cons of the mentioned methods.

Furthermore, risk analysis techniques can also be divided into the following two categories: deterministic methods and probabilistic methods.

> A deterministic method assumes that an unwanted event takes place and the physical effects and the damage of the event are calculated (hence, the event's probability of occurrence = 1 in such methods). On the contrary, a probabilistic approach takes both the likelihood and the outcome of an unwanted event into consideration.

Tab. 4.1: Suitable methods for engineering projects.

Project phase	Objectives	When	Required documentation	Analysis method
Conception	• Selection process • Identify unacceptable risks • Input to the design process • Identification of changes to reduce risks	Design evaluation	• Basic documentation	• PHA • Functional analysis • What-if • Brainstorming • Checklist • Project FMECA
Preliminary	• Identify the hazards associated with the process	• Design of the process • Flowcharts completed	• PFD* • Control flowchart • PID* • Process description	• What-if • Checklist • FMECA • HAZOP • FTA
Detailed engineering	• Identify hazards • Identify exploitation difficulties • Provide information for operation processes, design modifications, activation and maintenance.	Detailed engineering	• Final PFD* • Final control logic • Final PID* • Process description • Supplier drawings • Operation information	• Checklist • HAZOP • FMECA
Construction	• Verify conformity to plans and estimates during construction and equipment installation.	During construction	• PID* • Isometric design • Mechanical design • Norms and specifications • Supplier drawings	• Workplace inspection • Leak testing • X-ray sealing inspection • Non-destructive testing
Equipment hand-over	• Verify conformity of installation upon completion	End of construction	• PID* • Isometric design • Mechanical design • Norms and specifications • Supplier drawings	• Physical audit

(*Continued*)

(*Continued*)

Project phase	Objectives	When	Required documentation	Analysis method
Pre-operation	• Verify that equipment and process function as described	After equipment hand-over	• List of pre-operational tests • Functional description • Supplier information	• Dynamic tests with inert substances and controlled quantities of reactive substances
Before start-up	• Verify that the production system is safe before introducing chemicals	Before operational tests	• Risk analysis • HAZOP report • Training • List of deficiencies	• Plant inspection • Checklist
Pre-production	• Verify functionality of production system	After pre-activation review	• Report of pre-activation review • Corrective actions	• Dynamic tests with chemical substances

*PID, piping and instrumentation diagram; PFD, process flow diagram.

Tab. 4.2: Summary of the main characteristics of the risk analysis techniques.

Techniques	Procedure	Advantages	Disadvantages
FMECA	Examine whether components or process can have some failures	Good for equipment, mechanic systems	Little attention given to human factors Does not estimate cost of failure
HAZOP	Use the nodes of industrial plants to search for deviations from designed intent	Improve chemical process and operability	Time-consuming Experienced team leader required
ETA	Structuring cause back to the consequences	Quantitative with graphic tool Good for technology performing	Cannot analyze multiples failures
FTA	Structuring consequence back to the causes	Reveals the main causes of failure Give graphical view	Problem of reliability when data are minimized
RADM	Combining probability and severity of hazard. Determining a risk priority number	Graphical tool Good relation between probability and severity Ranking risks	Inadequate if there are many risks Cannot be used to deduce causes and consequences

(*Continued*)

(Continued)

Techniques	Procedure	Advantages	Disadvantages
PHA	Ask questions about potential failure, fault	Prioritize recommendations	Cannot be used to find details concerning a hazard
What-if	Checks for potential hazards by posing "What-if" questions	Very fast in searching for consequences	Cannot determine causes Very basic
Checklist	Use a list of hazards to record consequences and safety actions	Useful to have an overview of the hazards list	Much time required to find a hazards list
HRA	Evaluates human-machine interface, carry out task analysis.	Can help reducing human errors by improving performance shaping factors	Much time required if there are a lot of personnel

FMECA, failure mode, effect and criticality analysis; HAZOP, hazard and operability studies; ETA, event tree analysis; FTA, fault tree analysis; RADM, risk assessment decision matrix; PHA, preliminary hazard analysis; HRA, human reliability analysis.

A further distinction can be made according to the way in which a method is used:

- A qualitative approach: likelihood and consequences are treated purely in a qualitative way.
- A quantitative approach: both likelihood and consequences are fully quantified.
- A semi-quantitative approach (or semi-qualitative approach): both likelihood and consequences are quantified to a certain extent (within certain predefined limits).

Methods used according to a purely qualitative approach are generally less complex and are based on the use of arbitrary definable evaluation standards. Likelihood and potential consequences are described in detail. Such methods are rather simple, easy-to-use and flexible in their use, and can be applied to nearly all situations. The disadvantage consists in their dependence on subjective impressions and perceptions and in the fact that not all elements are taken into account.

Methods used following a quantitative approach try to structure events and situations in a systematic manner. A variety of scenarios and cause-consequence events are analyzed and the relevant parameters are identified. For every cause-consequence event the likelihood (under the form of a frequency or a probability) and the consequences are quantitatively determined. The result of such a study thus strongly depends on the reliability of the used values and data and the validity of the used models for the method. Such an approach is generally much more complex and more time-consuming than the qualitative approach.

Semi-quantitative approaches use detailed descriptions of likelihood and consequences and assign values to them according to the definitions provided. However, compared with a quantitative approach, the values are indicative and have much less a statistical background.

In ►Tab. 4.3 an overview is given of the advantages and disadvantages of quantitative and qualitative approaches. Semi-quantitative approaches are situated in between.

4.1.3 General procedure

All risk analyses are based on an identical procedure, usually iterative, that continues as long as the considered system does not reach an acceptable level of risk.

In reality, a risk analysis rarely crosses two-to-three iterations, except for systems where particularly critical safety and reliability are necessary (nuclear stations, spacecraft, etc.).

Tab. 4.3: Comparison between quantitative and qualitative risk analysis approaches.

	Consequence based	Likelihood based	Risk-based
Qualitative approach	Estimation of the consequences (no calculations)	Estimation of the likelihood (non-numerical basis)	Estimation of consequences and likelihood
	Advantages: • Quick • Easy-to-use to compare certain parameters, e.g., type of product, etc.	Advantages: • Quick • Easy-to-use to compare certain parameters, e.g., number of hazardous transports	Advantages: • Both consequences and likelihood are taken into account, thus the risk can be compared between specific parameters
	Disadvantages: • Likelihood is not taken into account • Provides no idea of the total risk, only an indication of the potential consequences • Cannot be applied for all applications/situations • Possibly not all relevant factors are considered • Provides only an estimation of the consequences usually based on a subjective opinion (expert judgment) – this entails the possibility that estimation by other experts leads to different results	Disadvantages: • Consequences are not taken into account • Provides no idea of the total risk, only an indication of the potential likelihood • Cannot be applied for all applications/situations • Possibly not all relevant factors are considered • Provides only an estimation of the likelihood usually based on a subjective opinion (Expert Judgment). This entails the possibility that estimation by other experts leads to different results.	Disadvantages: • Indicative consideration of risk, usually based on a subjective opinion (expert judgment). This entails the possibility that estimation by other experts leads to different results. • Cannot be applied for all applications/situations • Mainly for "rough" analyses, and not for detailed analyses

(Continued)

(*Continued*)

	Consequence based	Likelihood based	Risk-based
Quantitative approach	Calculation of the consequences	Calculation of the likelihood, based on causistic, historic data, and statistics	Both the consequences are calculated and the likelihood is estimated ($\rightarrow$ QRA, see also Section 4.9)
	Advantages: • Provides a scientific idea of the potential consequences based on objective parameters. • Results are less dependent on the person carrying out the risk analysis • Can more easily be used to suggest or to use acceptance criteria	Advantages: • Provides a scientific idea of the potential likelihood based on objective parameters • Results are less dependent on the person carrying out the risk analysis • Can more easily be used to suggest or to use acceptance criteria	Advantages: • Provides the most accurate picture of the risks • May be used to use acceptance criteria (if applied in an adequate way)
	Disadvantages: • More time-consuming than (semi-) qualitative method • Certain software and modeling required • Requires experience to handle the software and the models • Likelihood is not taken into account • Provides no picture of the total risk, only a calculation of the potential consequences • Detailed data required such as detailed environment factors, detailed weather conditions, topography, soil condition, etc. to obtain an accurate calculation	Disadvantages: • More time-consuming than in case of (semi-) qualitative approach • More data and analysis required than in case of (semi-) qualitative approach • More knowledge and experience required than in case of (semi-) qualitative approach	Disadvantages: • Most time-consuming approach • Certain software and models required • Experience required to handle software and models • A lot of background data and detailed information required

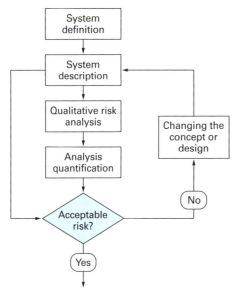

Fig. 4.2: General procedure of risk analysis.

The different steps are summarized in ▶Fig. 4.2. The essential starting point is the definition of the system. At this moment, a qualitative risk analysis can be undertaken. This analysis cannot always be quantified by figures or probabilities, often because of the lack of reliable and pertinent data. However, whenever it is useful and feasible, the system will be quantified. The final question that remains to be asked is: "Is the risk acceptable?" If yes, the analysis is complete, and if not, it becomes necessary to modify the concept or the design and repeat the analysis. This is the iteration of the process.

For both economic and efficiency reasons, it is generally favorable to integrate risk analysis at the project design stage (prospective analysis) (the measures are called "inherent" or "design-based" safety measures), because measures issued afterwards ("add-on" safety measures) are often more expensive and mainly palliative (like a bandage on a wooden leg).

▶Fig. 4.3 illustrates the evolution of costs linked to technical and operational modifications imposed by the results of risk analysis as well as their efficiency with time and development of the project.

4.1.4 General process for all analysis techniques

Risk assessment is an important step in protecting workers and business, as well as for complying with the law. It helps to focus on the risks that really matter in the workplace –

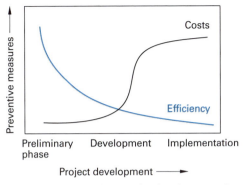

Fig. 4.3: Cost evolution depending on the development phases.

the ones with the highest potential to cause real harm, whether they are type I, II, or III risks. The general framework to assess risk follows five steps:

- Step 1: Identify the hazards.
- Step 2: Decide who or what might be harmed and how.
- Step 3: Evaluate the risks and decide on precautions.
- Step 4: Record your findings and implement them.
- Step 5: Review your assessment and update if necessary.

One of the crucial parts is the risk analysis procedure on which the decision-making process will be based. We can draft a global procedure that is valid for most available risk analysis methods. Some homemade techniques might also follow these rules; however if they are not openly accessible it is impossible to include them. The general risk analysis procedure could be divided into nine parts:

1. Definition of the system.
    - Objective(s) and scope of the study, definition of the system to be studied, identify the elements to be analyzed, subdivide complex processes.

2. Team selection.
    - Choose experts according to the process. Important factors are: multidisciplinary, expertise, availability.
    - Designate a secretary (generally the future user) or a moderator who will record the identified risks, causes, corrective measures, unsolved problems, etc.

3. Information gathering.
    - Collect all the necessary information before the analysis (products and equipment properties and description, operating procedures, technical drawings, process and flow diagrams, schemes, general drawings, process manuals, heat and mass flows, emergency procedures, weather conditions, environment, topography, human reliability, etc.).
    - Identify intended use.

4. Perform the analysis with the adequate chosen method.
    - Identify and list the elements to assess, make risk analysis meetings, save the results of the analysis in a table form or appropriate document, control the evaluation table by a system engineer, follow the methodology without introducing "feelings" statements, etc.
5. Recommend corrective actions and an action plan.
    - Define preventive and corrective solutions.
    - Recommend actions to reduce unacceptable risks.
    - Assign responsibility and schedule for corrective actions.
6. Monitor the implementation of the solution.
    - Regularly monitor the implementation of corrective measures.
    - Update the analysis in case of major changes.
7. Record hazards
    - Record identified hazards in the safety quality assurance system (if any).
    - Establish record keeping.
    - Establish documents of the complete analysis with diagrams, drawings, tables, processes.
    - Update information according to the completion of corrective measures.
8. Forecast to update the system
    - It must evolve to reflect changes in raw materials, formulation (recipe), market, habits or consumer demands, new hazards, scientific information, or inefficiency.
    - It must provide at the outset why, when and how the system will be reviewed.
9. Continuous monitoring and follow-up
    - Once the analysis is completed, the story does not stop there, as time is a factor of change, iteration of the procedure must be performed when (sometimes minor, certainly major) changes happen.

4.2 SWOT

SWOT stands for strengths, weaknesses, opportunities and threats. Strictly speaking, it is not a risk analysis technique. It is a widely used framework for organizing and using data and information gained from a situation analysis. It encompasses both internal and external environments. It is one of the most effective tools in the analysis of environmental data and information. However it allows for assessing strategic risks as the other techniques do not.

A SWOT analysis helps to identify, in a systematic and organized way, internal strengths and/or weaknesses. It helps with matching them with the opportunities or threats in the environment. SWOT is a business or strategic planning technique used to summarize the key components of the strategic environments. The three strategic environments are:

- Internal environment
- Industry environment
- Macro environment

Sometimes the last two are combined and are called "the external environment". The method of SWOT analysis is to take the information from an environmental analysis and separate it into internal (strengths and weaknesses) and external issues (opportunities and threats). Once this is completed, a SWOT analysis determines what may assist the firm in accomplishing its objectives, and what obstacles must be overcome or minimized to achieve desired results.

Factors affecting an organization can usually be classified as in ▶Fig. 4.4:

SWOT is very often represented as a matrix to allow comparing for all the sensitive aspects in one shot (▶Fig. 4.5).

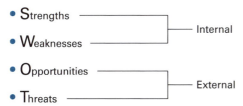

Fig. 4.4: Internal and external environment.

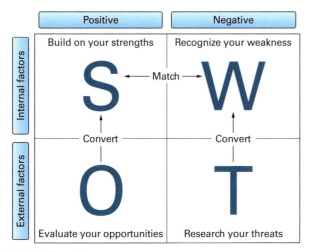

Fig. 4.5: The SWOT matrix.

Effectively, one has to match each component with another. For example, match the internal strengths with external opportunities and list the resulting strengths/opportunities strategies in the matrix chart. The four strategy types are:

- S–O strategies pursue opportunities that match the company's strengths. These are the best strategies to employ, but many firms are not in a position to do so. Companies will generally pursue one or several of the other three strategies first to be able to apply S–O strategies.
- W–O strategies overcome weaknesses to pursue opportunities. Match internal weaknesses with external opportunities and list the resulting W–O strategies.
- S–T strategies identify ways that the company can use its strengths to reduce its vulnerability to external threats. Match internal strengths with external threats and list the resulting S–T strategies.
- W–T strategies establish a defensive plan to prevent the firm's weaknesses from making it susceptible to external threats. Match the internal weaknesses with external threats and record the resulting W–T Strategies.

In order to fill this matrix, several questions must be answered related to the four aspects of SWOT as depicted in ▶Fig. 4.6 (some illustrative questions are presented).

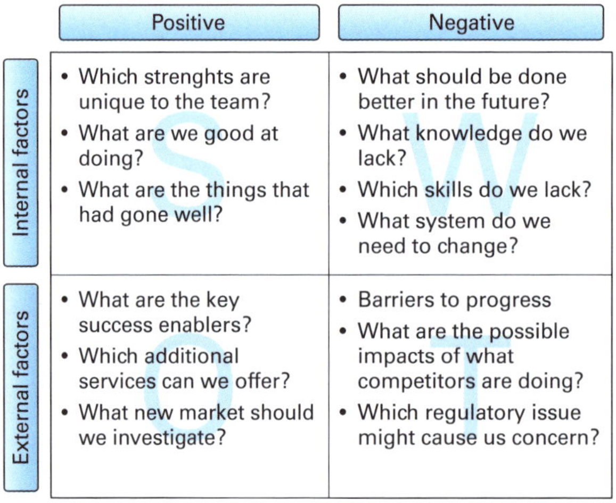

	Positive	Negative
Internal factors	• Which strenghts are unique to the team? • What are we good at doing? • What are the things that had gone well?	• What should be done better in the future? • What knowledge do we lack? • Which skills do we lack? • What system do we need to change?
External factors	• What are the key success enablers? • Which additional services can we offer? • What new market should we investigate?	• Barriers to progress • What are the possible impacts of what competitors are doing? • Which regulatory issue might cause us concern?

Fig. 4.6: The SWOT sample questions.

As stated in the beginning, a SWOT analysis will summarize your strategic analysis. It will be an integral part of the four-step planning process as illustrated in ▶Fig. 4.7.

In conclusion, a SWOT analysis is a useful technique for understanding your strengths and weaknesses, and for identifying both the opportunities open to you and the threats you face. Used in a business context, a SWOT analysis helps you carve a sustainable niche in your market. Used in a personal context, it helps you develop your career in a way that takes best advantage of your talents, abilities and opportunities.

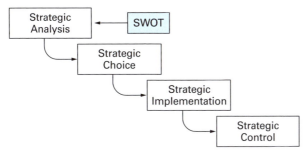

Fig. 4.7: A link to strategy.

4.3 Preliminary hazard analysis

Preliminary hazard analysis (PHA), is a relatively simple technique to implement and it allows a quick identification of the main risks of a system. PHA was instituted and promulgated by the developers of the US Air Force standard practice for system safety (MIL-STD-882) since 1969 (3). PHA is often used in the early life of the processes (conceptual phase or research and development phase) (CCPS 2008) (4) to affect the design for safety as early as possible. The technique is a safety analysis tool for identifying hazards, their associated causal factors, effects, level of risk, and mitigating design measures when detailed design information is not available. To perform the PHA analysis, the system safety analyst must have three information inputs – design knowledge, hazard knowledge, and a preliminary hazard list (collection of identified hazards). The output of a PHA includes identified and suspected hazards, hazard causal factors, the resulting mishap effect, mishap risk, safety critical function, and top level mishaps. The advantages of PHA are: that it is easily and quickly performed; it is comparatively inexpensive in providing meaningful results; it is a methodological analysis technique; most of the system hazards are identified and an indication of system risk is provided. While there are no major disadvantages to the PHA, there is sometimes an (improper) tendency to have it as the only applied analysis technique (3).

A PHA is applied in two ways:

1. Alone, as risk analysis for systems with simple or easily identifiable hazards and not complex accidental process.
2. In combination with other methods. In this case, a PHA is seen as a preliminary risk study to prepare the complex or poorly defined case. In this sense, it is mainly used in the early design phase of a project.

PHA is an identification and analysis technique of the hazard frequency that can be used in preliminary conception phases in order to identify hazards and evaluate their criticality. However, usually PHA is not limited to the risk evaluation phase, but also gives a certain directive in order to master these risks, making it a management method rather that a risk analysis.

PHA aims at listing all sources of hazards present in the system: hazardous materials, equipment or components that could present a hazard, risky processes or procedures. Each of these elements will be assigned to one or more adverse events, possible causes, as well as potential compensatory measures. Results are generally presented in a table. It requires, as a first step, to identify the hazardous parts of the installation/process. These harmful elements refer most often to:

- Hazardous substances and preparations, whether in the form of raw materials, finished goods, utilities, etc.
- Hazardous equipments such as storage facilities, reception areas, shipping, reactors, energy supply (boilers, etc.)
- Hazardous operations associated with the process.

An example of a PHA applied to a storage tank is presented in ▶Tab. 4.4.

The main advantage of the preliminary hazard analysis is to enable a relatively quick review of hazardous situations on the facilities. Relatively economical in terms of time and resources, it does not require a very detailed level of description of the studied system (generally implemented at the design stage). This method is

Tab. 4.4: Example of PHA analysis on storage of flammable gas under pressure.

Element	Hazard	Hazardous event	Causes	Consequences	Measures
Tank	Heat stress (fire outside the tank)	Explosion of the tank and important release of gases	Presence of combustibles elements near the tank	• Fire • Property damage • Casualties	• Changing logistics storage • Move individual hazards away
Tank	Mechanical impact against the shell of the tank	Gas release	• Accident with a crossing vehicle • intentional damage	• Fire • Property damage • Casualties	• Inspection program • Continuous monitoring of air quality
Tank	Weakening of the tank shell	• Gas release • Explosion of the tank and important release of gases	• Corrosion • Fatigue (crack) • Not well sized tank (do not stand up to pressure imposed)	• Fire • Property damage • Casualties	• Inspection program • Verification of design
Valve	Unexpected opening	Gas release	• Valve or control system failed • Error during routine maintenance	Gas release	• Inspection program • Continuous monitoring of air quality

a simplified form of the FMEA, but it is very limited in failures propagation and consequences of multiple failures. Only the direct consequences of failure are known. Moreover, this method does not require a modeling of the system, it is difficult to ensure a systematic approach, and therefore it is based on the knowledge of the team conducting the analysis. PHA must therefore really be considered as a preliminary method or a sufficient method for systems simple enough to be appropriately studied without modeling.

> Preliminary hazard analysis (PHA) is a qualitative analysis that is performed to:
> - Identify all potential hazards and accidental events that may lead to an accident.
> - Rank the identified accidental events according to their severity.
> - Identify required hazard controls and follow-up actions.
>
> Several variants of PHA are used, and sometimes under different names, such as: rapid risk ranking or hazard identification (HAZID).

4.4 Checklist

A checklist analysis uses a written list of items or procedural steps to verify the status of a system (5). Traditional checklists vary widely in their level of detail and are frequently used to indicate the compliance with standards and practices. The checklist analysis approach is user-friendly and can be applied at any stage in the lifetime of the process. Even though it will not replace a detailed risk analysis process, a checklist is a good starting point and a highly cost-effective method for common hazards.

The checklist is the easiest to implement, but remains very limited in what it can offer. Either it is too generic and could be difficult to be applied to the system, or it could be considered as appropriate to the system, which implies in this case that hazards have been previously identified.

The checklist uses a standard form containing a series of specific questions related to the potential hazard of the considered process.

4.4.1 Methodology

The checklist involves the use of a pre-established form containing a series of specific questions related to the potential hazard of the considered process. A control list, or checklist, is also an operation involving the methodical verification of the necessary steps that are required for the process to run with a maximum of safety. An example would be the use of such a procedure in aviation. It becomes a safety and security procedure that methodically verifies whether the plane is ready to undergo the next phase of flight. Such operations are usually carried out vocally or by ticking a written procedure list. The principle of hazard determination and action planning with the use of a checklist is presented in ▶Fig. 4.8.

Identify hazards 	1	What are the risks to **health and safety** at the workplace? The checklists can help you **identifying hazards**.
Take the **necessary** measures 	2	Plan and implement the **appropriate safety** measures. Checklists and other publications provide appropriate safety measures.
Act methodically 	3	Important is to act methodically! Proceed methodically to ensure safety in the long term business.

Fig. 4.8: Principle of hazard identification and action planning using a checklist.

Its principle is based on the operating procedure structured in numbered steps, each corresponding to a simple operation. Every step of the process is evaluated in terms of failures that could occur and possible deviations from normal operating procedure. This widely used method can be adapted to many situations. Several organizations have developed checklists for different activities (e.g., Suva in Switzerland, www.suva.ch).

The methodology takes place in three steps:

Each element for a given step, or whose failure represents a risk, will be marked with a cross, indicating a potential hazard.

Each marked hazard will be deferred and described in detail as a script, so as to assess its likelihood of occurrence and severity.

Measures will be implemented in order to reduce risks for intolerable situations.

4.4.2 Example

Let us assume that we have divided a process into several steps, each identified with a reference number. In the table (see ▶Tab. 4.5 and 4.6) we identify by a cross mark the process steps that are sensitive to the different failures possibilities.

4.4.2.1 Step 1a: Critical difference, effect of energies failures

In the example displayed in ▶Tab. 4.5, point 12 (whether it refers to a stage, a unit operation or a process step) is liable to risk in case of electricity or compressed air regulation failures.

We then move forward by implementing the same procedure but this time originated by possible deviations from the operating procedure. The question to be answered is "Which process steps might be affected when an operating parameter could deviate from a desired value or function?" One can observe in ▶Tab. 4.6 that step no. 18 is sensitive to ventilation as no. 75 is liable to risk when the set temperature is no longer what is expected.

Tab. 4.5: Process steps sensitive to energies failures.

Unit operation/Step number	12	43	54	68	75
Electricity	x				x
Water		x			
Vapor			x		
Brine/ice					
Nitrogen					
Compressed air (regulation)	x				
Compressed air (command)					
Vacuum				x	
Ventilation			x		
Absorption					

4.4.2.2 Step 1b: Critical difference, deviation from the operating procedure

Tab. 4.6: Process steps sensitive operating procedure deviations.

Unit operation/Step number	14	18	55	74	75
Cleaning	x				
Plant inspection					
Discharge					
Equipment ventilation		x			
Charges, dosage					
Quantity, flow rate	x				
Operation succession			x		
Speed of addition	x				
Product mismatch					
Electrostatic charges					
Temperature					x
Pressure	x				
pH				x	
Heating/cooling					
Speed of agitation					
Reaction with coolant					
Catalyst, inhibitor					
Connecting lines, valves					
Process interruption			x		

4.4.2.3 Step 2: Establish the risk catalogue

For the identified critical steps (defined by the hazard description) one may evaluate the risk by identifying the likelihood of occurrence and the severity. Then mitigation measures have to be identified and proposed to reduce the risk. The first row of the table (see ▶Tab. 4.7)

Tab. 4.7: Extract of an example of a risk catalogue.

Product: Plant: Author:	No. identity: Bldg:		Process: Date:		Revision:
Hazard description (causes, consequences)	Evaluation		Description of (technical) risk mitigation measures	Evaluation residual risk	
	O	S		O	S
Overheating of the reactor R-321 at the step 75 (synthesis step)	F	H	Temperature control, alarm high, cutting feed stream F-32	F	F
No more control system at the step 12, loading the reactor, in the absence of electric current	F	H	UPS system for the automatic control	F	M
Failure of ventilation at step 18, product loading TX24	M	H	No technical measures possible	M	H

O, likelihood of occurrence; S, Severity;   F, low; M, medium; H, high.

helps to identify the process, product, plant building, the author, date and revision number (when successive analysis was performed). In the example we note that step 75 reveals a hazard that may lead to the overheating of the reactor R-321. It has been marked "low" for the occurrence and "high" for the severity. The following mitigation measure was proposed: setting a temperature control with a high alarm level setting the stoppage of the steam feed F-32. The authors of this example then evaluate both occurrence and severity as "low" when talking about residual risk (the risk after corrective measures have been taken).

4.4.2.4 Step 3: risk mitigation

Measures to reduce the risk for each hazard will be clearly defined so their descriptions can be used as specifications. We then evaluate the progress by assessing the residual risk.

4.4.3 Conclusion

Although this qualitative method is limited in discovering hazards that have not already been identified, it can still prove to be efficient when applied to a new system, with similar functions to an existing one for which the checklist was established. Furthermore, the checklist can be easily used by relatively inexperienced personnel, provided that the list has been established by experts. In this case the checklist allows the user, e.g., to confirm that the process design does not exhibit safety vulnerabilities. The checklist is not adapted to new technologies whose hazards are insufficiently recognized, identified and understood.

> Checklist analysis is a systematic evaluation against pre-established criteria in the form of one or more checklists. It is a systematic approach built on the historical knowledge included in checklist questions. It is used for high-level or detailed analysis, including root cause analysis. It is applicable to any activity or system, including equipment issues and human factors issues. It is generally performed by an individual trained to understand the checklist questions. It generates qualitative lists of conformance and non-conformance determinations, with recommendations for correcting non-conformances. The quality of evaluation is determined primarily by the experience of people creating the checklists and the training of the checklist users.

4.5 HAZOP

The HAZard and OPerability study (HAZOP) was developed by Imperial Chemical Industries (ICI) in the late 1960s and the earliest work from ICI Mond Division in the northwest of England was published in 1968 (6). The HAZOP system as we know it today was published by Lawley from ICI Petrochemicals Division in the northeast of England in 1974. After the Flixborough disaster in 1974 (7), this technique became widely used.

HAZOP is an organized methodological technique for analyzing hazards and operational concerns of a system, often used in chemical industries (8–12). According to HAZOP, normal and standard operations are safe and hazards occur only when there is a deviation from the normal operation (3).

The standard CEI 61882 defines the objectives of the original HAZOP method as:

- Identification of potential hazards in the system. The hazard can be limited to the immediate proximity of the system or spread its effects beyond the system, such as environmental hazards.
- Identification of potential exploitation problems of the system and particularly their causes, identification of functional perturbations and deviations in production liable to lead to the production of non-complying products.

With the introduction of the directive Seveso II in Europe and new requirements regarding the prevention of industrial risks, the original HAZOP method became insufficient for the analysis of major risks. A phase of risk evaluation was added to the original method, and the previously purely qualitative HAZOP method became semi-quantitative, contributing to the improvement of the knowledge of risk and thus the safety of facilities.

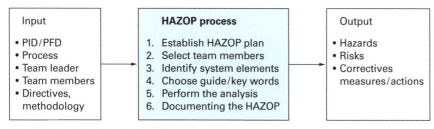

Fig. 4.9: Principle of the HAZOP process.

4.5.1 HAZOP inputs and outputs

HAZOP's inputs are design data, guidelines and process description (see ▶Fig. 4.9). In order to perform systematic searches for conceivable departures from the design intent [normally presented in a piping and instrumentation diagram (P&ID)], the technique uses guide keywords (more, none, less, etc.) combined with process/system conditions such as speed, flow, pressure, and temperature. P&ID are scrutinized vessel by vessel, pipe by pipe, to ensure that all potentially hazardous situations have been taken into account. Once the hazard resulting from potential deviation in design operation is identified, a search backwards is performed to find possible causes and forward to find possible consequences (middle-bottom-up technique).

HAZOP's outputs are therefore: hazard, cause, consequence, and corrective measures (see ▶Fig. 4.9).

4.5.2 HAZOP process

HAZOP allows, in an inductive process, to systematically analyze the deviations of the plant or process parameters and to predict their potential consequences. This method is particularly useful for the examination of thermo-hydraulic systems, for which parameters such as speed, temperature, pressure, level, concentration, etc. are particularly important for the safety of the installation.

By its nature, this method requires particular consideration of schemes and plans of fluid flow patterns or P&ID and process flow diagram (PFD). However, the method can be applied and elaborated for virtually any system.

Based on a detailed description of the process, each part of the facility will be analyzed as to possible deviations from normal operation. Causes, effects or consequences and remedies are sought. It requires a team of experts, each with special skills and competences and a moderator or facilitator controlling the method.

The method could be expressed in five steps (see also ▶Fig. 4.10):

1. Subdivision of the process

    - Divide the process into unit operations with their auxiliary instruments.

2. Expected function

    - Scope of the unit operation, intention description.

3. Search for deviations

    - Each unit operation and ancillary facilities will be reviewed in light of possible deviations from the targets using the guide's word.
    - Deviation = Guide/keyword + parameter
    - Identify the possible causes and consequences of those deviations.
    - Repeat the search for deviations to each unit. The process then stops when all the units and sub-units have been studied.

4. Risk assessment

    - Assessing the likelihood of occurrence and severity.
    - Comprehensive analysis of all conceivable realistic deviations.

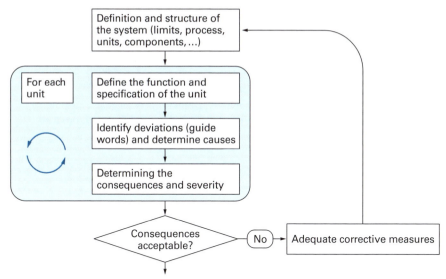

Fig. 4.10: Methodology of HAZOP.

5. Measures
   - For consequences deemed too serious, measures must be proposed to eliminate the cause, or to reduce the severity or impact.

In step 3, for each constituent part of the system considered (line or mesh); a deviation generation (conceptual) is performed systematically by the conjunction of:

guide/keywords (7 technical + 4 temporal) see ▶Tab. 4.8 and parameters associated with the system studied, see ▶Tab. 4.9. The parameters commonly encountered are temperature, pressure, flow, concentration but also the time or the operations to perform.

How to conduct the analysis?

1. Initially, choose a line of process. It generally covers equipment and connections, all performing a function in the process identified in the functional description.
2. Choose an operating parameter.
3. Identify a guide/keyword and generate a deviation.
4. Verify that the deviation is credible. If yes, proceed to step 5, otherwise return to step 3.
5. Identify causes and potential consequences of this deviation.
6. Examine ways to detect this drift as well as those provided to prevent the occurrence or mitigate its effects.
7. Propose, where appropriate, recommendations and improvements (see ▶Tab. 4.10).
8. Choose a new guide/keyword for the same parameter and return to step 3.
9. When all guide/keywords have been considered, choose another operating parameter and go back to step 2.
10. When all operating phases have been studied, choose another process line and go back to step 1.

Tab. 4.8: HAZOP guidewords or keywords.

Keywords	Signification	Commentary	Examples
No or not	No part of the intention is fulfilled	The purpose or function is not fulfilled at all, not even partially	• No agitation • No flow
More	Overrun, or increase, quantitatively	Refers to the quantities and properties (T, P), but also activities (heating, reaction)	• Higher temperature • Too much product
Less	Insufficient or quantitative reduction		• Lower flow rate than expected • Less agitation
As well as	Qualitative increase	• The intent (design and procedure) is performed with additional activity • Concomitant adverse effect	Heating started at the same time as the addition of reagent A
Part of	Qualitative modification/diminution	Only part of the intention is realized	Only part of the reagent is added
Reverse	The logical opposite of the intention	Reversal of the activity or sequence	• Liquid flows in the opposite direction • It heats instead of cooling
Other than	Total substitution	Result different from that of intention	Reagent A is loaded in place of B
Earlier than	On the time clock	The action takes place before or after a defined time	We started heating 15 minutes before the deadline
Later than	On the time clock		The reaction, taking place over 2 hours, has crossed the deadline by 1 hour 45 minutes
Before	On the sequence or order	Action is taken before or after the defined sequence	A was loaded before B
Later	On the sequence or order		It was cooled after stirring

Tab. 4.9: Example of operating parameters.

Measurable physical quantities		Operations		Actions	Functions-situations
Temperature	pH	Loading	Control	Start-up	Protection
Pressure	Intensity	Dilution	Separation	Sampling	Utility default
Level	Speed	Heating	Cooling	Stop	Freezing
Flow rate	Frequency	Stirring	Transfer	Isolate	Spill
Concentration	Amount	Mixing	Maintenance	Purge	Earthquake
Contamination	Time	Reaction	Corrosion	Close	Malevolence

Tab. 4.10: Example of safety barriers.

Safety Barriers		Definition	Example
Technical	Passive safety devices	Unitary elements aiming to fulfill a safety function without external energy supply from the system to which they belong, and without the involvement of any mechanical system	• Holding tank/tray • Rupture disc
	Active safety devices	Items not passive aimed to perform a safety function without an external energy supply system to which it belongs	• Safety valve • Excess flow valve
	Safety instrumented systems	Combination of sensors, processing units and terminal elements aiming to fulfill a function or a sub-safety function	• Measuring elements that controls a valve or switch power
Organizational		Human activities (operations) that do not involve technical safety barriers to oppose the conduct of an accident	• Emergency plan • Containment
Systems with manual action		Interface between a technical barrier and human activity to carry out a safety function.	• Pressing an emergency button • Low flow alarm, followed by manual closing of a safety valve

Tab. 4.11: Example of HAZOP analysis table.

Phase, function: (detailed description, notated)										
Guide words: (examples) M1: no or not M2: less M3: more	M4: As well as M5: Part of M6: Reverse M7: Other than	M8: Lather than M9: Earlier than M10: Before M11: Later				Level of P and G: (L)ow, (M)iddle, (H)igh P1, G1 occurrence and severity before measures P2, G2, after measures				
Guide word	Deviation	Possible cause	P1	Consequences	G1	Measures	P2	G2	Who	When
1	2	3	4	5	6	7	8	9	10	11

4.5.3 Example

Results of the analysis are often represented in a table (see ▶Tab. 4.11), indicating the function or the phase evaluated (0), the keyword used (1), the observed deviation (2), identification of the possible causes (3), the evaluation of the occurrence and severity (4 and 6), which are the consequences (5), the correctives measures to be applied (7), the re-evaluation of the occurrence and probability (8 and 9), who is responsible to apply the measures (10) and the deadline for implementation (11).

Let us take an example that deals with the water supply to a cooling system. Results of the analysis are presented in ▶Tab. 4.12, revealing that parameters fluid and electricity are the most critical. We intentionally took another representation, indicating that there is no universal result table. It is important that all information is present.

4.5.4 Conclusions

The HAZOP method is systematic and methodical. Considering simple parameter deviations of the system, it avoids among others to consider, such as FMEA, all possible failure modes for each component of the system. In contrast, the HAZOP makes it difficult to analyze the events resulting from the combination of multiple simultaneous failures. Furthermore, it is sometimes difficult to assign a guideword/keyword to a portion of well-defined system to study. This complicates the comprehensive identification of potential causes of deviation. The studied systems are often composed of interconnected parts so that a deviation occurring in a line or mesh can affect causes or consequences on a neighboring mesh and vice versa. It is possible a priori to see the implications of a deviation from one part, to another system. However, this task can quickly become complicated. Finally, HAZOP makes it difficult to analyze the events resulting from the combination of multiple simultaneous failures.

HAZOP has several advantages. It is easily learned and clearly structured. It provides rigor for focusing on system elements and hazards; it is a team effort with many viewpoints. The introduction of modeling could help, e.g., with taking into account the discrete components in the system; their connection and behavior over time (13,14).

HAZOP has also some inconveniences:

- HAZOP analysis focuses on single events rather than combinations of possible events.
- It is based on guide words, allowing us to overlook some hazards not directly related to them.
- HAZOP analysis training is essential for optimum results, especially for the facilitator.
- HAZOP analysis can be time-consuming and thus expensive (3). It is not uncommon that over 25 man-days' work is necessary to perform an HAZOP study. Of course, the time depends on the scope (e.g., complexity of an installation) of a study.

> Hazard and Operability Analysis is a structured and systematic technique for system examination and risk management. In particular, it is often used as a technique for identifying potential hazards in a system and identifying operability problems that are likely to lead to nonconforming products. It is based on a theory that assumes risk events are caused by deviations from design or operating intentions. Identification of such deviations is facilitated by using sets of "guide words" as a systematic list of deviation perspectives.

What-if, an inductive method similar to HAZOP (although much less systematic and more intuitive), is a brainstorming approach in which a group of experienced

Tab. 4.12: Partial result of an HAZOP analysis.

N°	Object	Function	Parameter	Key word	Consequence	Cause	Hazard	Risk P/G	Recommendation	Risk 2 P/G	Who	When
1	Line	Bring water to the system	Fluid	No	Loss of cooling of the pump	Line rupture	Damaged pump	LM	–	LM	–	–
2				More	Pressure rise in the line	No pressure regulation	Line rupture	LM	Add a safety valve to the loop	LL	TJ	2012.10
3				Less	Not enough cooling capacity of the pump	Leakage at the pipe and fittings	Damaged pump	MH	Periodic control of fittings	LM	TJ	Bimonthly
4	Electrical supply	Supply power to the motor M23	Electricity	No	Loss of power to the pump	Short circuit, power failure	Loss of control	LH	Backup Power	LL	JK	2013.01

people familiar with the subject process raise the question "what-if?" instead of using keywords when examining the P&ID and voice concerns about possible undesired events.

4.6 FMECA

The failure mode and effects analysis (FMEA) was developed by the US military in 1949 and furthermore encouraged in the 1960s in the aerospace industry. The Ford Motor Company reintroduced it in the late 1970s for safety and regulatory consideration and used it effectively for production and design improvement. Nowadays (15–18), this method is often used in industries producing machinery, motorcars, mechanical, and electronic components (19). A more detailed version of FMEA is called "failure mode, effects and criticality analysis" (FMECA), which adds prioritizations of actions to be taken, based on a risk score.

The terms used in FMECA are the following (see also Chapter 3):

- Reliability: The ability to present no default for a specified period under specified conditions.
- Availability: The ability to provide a given function under given conditions at a given time.
- Maintainability: The ability to be back to service in a given period under given conditions.
- Safety: The ability to present no hazard to people, property and the environment.
- Failure: The termination of the ability of an entity to perform a required function. Element, state action no longer fulfilling its original function, expected or anticipated.
- Failure mode: A potential failure mode describes the manner in which a product or process could fail in fulfilling its primary function.
- Failure cause: The process or mechanism responsible for the initiation of failure mode.
- Failure consequence: The result that the failure mode induced on the operation or function.
- Detection: An assessment of the likelihood that the controls (design and process) will detect the failure cause or the failure itself.
- Control: Controls (design and process) are the mechanisms preventing the cause of a failure to occur.

A failure occurs between a cause and an effect. A single cause may have multiple effects. A combination of causes could lead to a single or multiple effects. Causes can themselves have causes and effects may have subsequent following effects (see ▶Fig. 4.11).

FMEA is an inductive method based on the study of elementary failures of composed systems in order to deduce what they can result in and therefore what situations can be expected due to these failures. FMECA includes an additional evaluation, which studies the severity in these situations. It includes identifying and evaluating the impact of elementary failures on the corresponding system, functions and environment. FMEA and FMECA are so commonly used and well-known that they have practically become the safety symbol of functional safety. As FMEA is included in FMECA, we will refer to both as FMECA in this chapter.

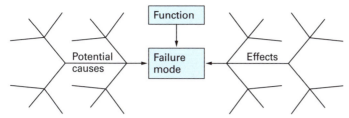

Fig. 4.11: Cause-effect failure.

4.6.1 FMECA inputs and outputs

FMECA is a tool for inductive and quantitative analysis based on the study of possible failure modes of components and functions constituting a system. It is a powerful method suitable for complex or innovative systems with substantial implementation. It is well-adapted to systematically and thoroughly identify all undesirable situations which may lead to individual failures.

FMEA evaluates the effects of potential failure modes of subsystems, assemblies, components and functions with design, functional diagrams and failure knowledge as input. Outputs of the technique are: failure modes, consequences, reliability prediction, hazards and risks, critical item list. See also ▶Fig. 4.12.

4.6.2 FMECA process

The global FMECA process is performed according the following procedures: i) System definition, ii) FMECA plan, iii) Team selection, iv) Collecting information, v) Perform FMECA, vi) Define corrective mitigation measures, vii) Monitor the implementation, viii) Register hazards and ix) Analysis documentation.

The implementation of the FMECA methodology follows three steps:

1. Develop a hierarchical model of the system in question.
2. Identify failure modes associated with each base unit model.
3. Identify the propagation of failures through the model and determination of final consequences.

The analysis and evaluation from the functional diagram of failure modes could be summarized as (see ▶Fig. 4.13):

- Look systematically for potential failure modes of components.
- Accurately describe the effect on the client.

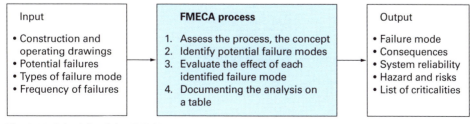

Fig. 4.12: Principle of the FMECA process.

116 — 4 Risk diagnostic and analysis

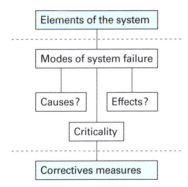

Fig. 4.13: Principle of FMECA process.

- List the possible causes of failure mode.
- List the validation of systems planned.
- Evaluate the modes, calculate the criticality index.
- Prioritize modes.

4.6.2.1 Step 1: Elaboration of the hierarchical model, functional analysis

It is important to note that the construction of the hierarchical structure is defined top-down, while the propagation of failures is achieved as inductive bottom-up (▶Fig. 4.14).

This first step is therefore the decomposition of the system into smaller elements (sub-systems). What should we obtain at this stage?

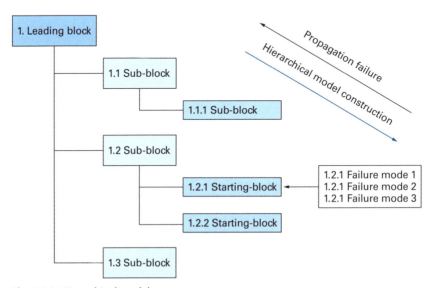

Fig. 4.14: Hierarchical model.

- The decomposed elements must be simple in order to completely identify their failure modes.
- The group must consist of an expert for each element who is capable of describing its nominal function and its failure modes in detail.
- The whole system must be covered through the decomposition.

It is not important for the decomposition to be finally separated but it is preferable for the various levels to be close enough to ensure that each component of the system should be present in the decomposition at least once (ideally only once). The level of decomposition is the most important decision in this step. The main purpose is to know why and how the studied product or process operates:

- Product decomposition into simple elements or subsets.
- The organizational system orders all the elements of the system on different hierarchical levels from the top.

As an illustration the decomposition of a grinder is shown in ▶Fig. 4.15.

This first step therefore results in a list of elementary functions or a list of components of elementary subsystems containing each function, sub-system or component affected by the different failure modes it may encounter. One must not overlook the fact that the failure modes of the components do not only depend on the component itself but also on the conditions in which they are implemented. This is the reason why the real experience of the enterprise or profession is always more valuable than data available in other places around the world. It is also important not to forget, neglect or discard certain failure modes because of their apparent weak frequency or probability of occurrence.

4.6.2.2 Step 2: Failure mode determination

The second step involves imagining and describing what would happen to the system when a studied failure mode appears. This is the stage where one may regret not having sufficiently decomposed the system. If each failure mode leads to a cascade of effects depending on various parameters (depending on the working stage, e.g.), the procedure would be difficult to carry out and it may be better to go one step back, define several FMECAs according to the same parameters and re-evaluate them one by one.

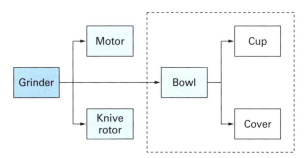

Fig. 4.15: Functional analysis of a grinder.

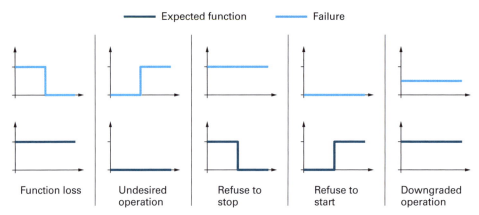

Fig. 4.16: Generic failure modes (top the failure and bottom the expected function).

They are five generic failures modes in ▶Fig. 4.16, going from the function loss, the undesired operation, the refusal to stop or to start, to the downgraded operation.

The following must be identified for each failure mode:

- Its causes (weighted based on probability or likelihood of occurrence).
- Its effects (weighted based on severity).
- Measures to counter or limit the effects of the failure (weighted based on the probability of non-detection).

One must describe the effects seen from outside of the system, the effects on the accomplishment of the function in the log, described by an external functional analysis; and also the effects of the system's safety or of the environment that are not written anywhere else. The concept is then based on the following questions:

- What can fail?
- How does it fail?
- How frequently will it fail?
- What are the effects of the failure?
- What is the reliability/safety consequence of the failure?

FMECA is a qualitative and quantitative method based on five questions (▶Fig. 4.17):

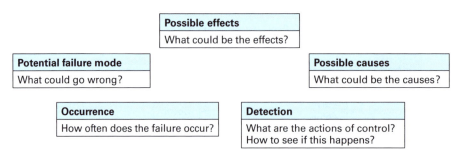

Fig. 4.17: Five key question of the FMECA analysis.

4.6 FMECA

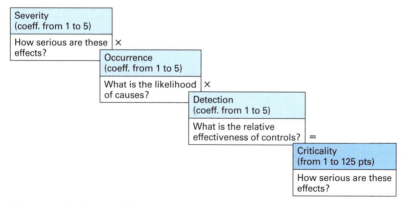

Fig. 4.18: Calculation of the criticality index.

4.6.2.3 Step 3: The criticality determination

The criticality or risk priority number is an indicator of the importance of the identified impacts, and therefore a synthesis of several parameters that signify the importance of failure modes. The choice of parameters depends on the subjects being analyzed and the results being sought. It is important to explain what the criticality is composed of (see ▶Fig. 4.18). A common approach is to undertake a synthesis of the severity of the consequences, the occurrence of the failure mode and its possibility of detection as follows:

Criticality = Severity × Occurrence × Non-detection

This is the synthesis of questioning and it allows for a prioritization of concerns. The risk priority index is used to help identify the greatest risks, leading to corrective action.

- The severity of the consequences is based on a scale of 4, 5 or 10 levels. It is a type of measure of consequences of various possible scenarios weighed by their conditional probability once the initial failure takes place.
- The occurrence of the failure mode is an evaluation of number of cases per unit of time. It is linearly represented on a scale containing as many levels as the severity scale.
- The detectability or non-detection is represented by a coefficient on a scale of 2, 3, 5… levels reflecting the severity of the consequences in case of detection weighed by the probability of detection. Failure modes are detected and stopped through:
  - Controls
  - Measurements
  - Calculations/Modeling
  - Procedures
  - Training

The main questions asked after criticality evaluation can be summarized in ▶Fig. 4.19.

4.6.3 Example

Results of the analysis are often summarized in tables (▶Tab. 4.13), like the HAZOP risk analysis, where each corrective measure is described with a list of all defects on which it has an impact (mitigation of occurrence or consequences, including when possible negative impacts).

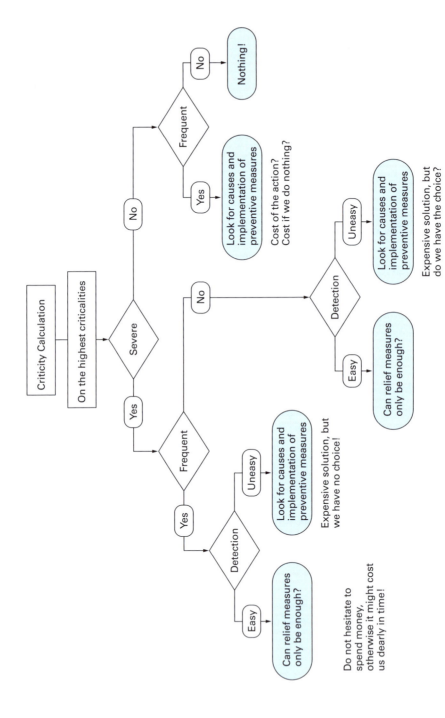

Fig. 4.19: Criticality calculation diagram.

4.6 FMECA

Tab. 4.13: Example of an FMECA table.

Process		Occupational safety		Estimation				Corrective measures	Estimation				
No	Function/activity	Hazards	Damage/injury	F	G	D	C	Measures	F	G	D	C	Date/visa
1	2	3	4	5	6	7	8	9	10	11	12	13	14

F, frequency; G, Severity; D, Non-detection; C: criticality index.

The following topics have to be determined: a reference number (1), the function or activity evaluated (2), the identified hazards (3), the potential damages or injuries (4), the estimation of the occurrence, severity, non-detection and calculation of the criticality index (5–8), the correctives measures to be applied (9), the re-evaluation of the criticality (10–13), who is responsible to apply the measures and the deadline for implementation (14).

The following example in ▶Tab. 4.14 depicts an extract of a FMECA analysis for a chemical lab.

Tab. 4.14: Example of FMECA analysis.

Process		Occupational safety		Estimation			
No	Function/activity	Hazards	Damage/injury	F	G	D	C
1	Gas bottle handling (>10L)	Fall of the cylinder	Foot crush up	3	2	5	30
2		Valve rupture after shock	Person crush up	1	5	5	25
3	Handling chemicals	Spillage on person	Contamination	3	2	5	30
4		Spillage	Contamination	4	5	5	100
5	Cryogenics handling	Spillage on person	Burn	2	2	5	20
6		Asphyxia	Death	1	5	5	25

Corrective measures	Estimation				
Measures	F	G	D	C	Date/Visa
Transportation chariot	1	2	5	10	2013.10/JM
Transportation only with protective cap	1	2	5	10	2012.07/TJ
Transportation with tray on chariot	1	2	5	10	2012.07/TJ
Transportation with tray on chariot	1	2	5	10	2012.07/TJ
Chariot ad hoc + gloves, goggles, lab coat	1	1	5	5	2012.11/JM
Oxygen detectors	1	2	1	2	2013.05/JB

> FMEA and FMECA are methodologies designed to identify potential failure modes for a product or process, to assess the risk associated with those failure modes, to rank the issues in terms of importance and to identify and carry out corrective actions to address the most serious concerns.

4.6.4 Conclusions

The analysis of failure modes and their effects with or without criticality is an approach based on logic and common sense. Considering that no system is infallible, this analysis includes identifying, describing and evaluating risks that result from failures. It is a pertinent method whenever the failure modes of components and the internal functioning of the system are or can be known and understood. There are limitations because of the method itself (not suited to represent the dynamics of a system, the temporal dimension and logical combinations) and the information available (like all methods of safety engineering, the existing and available information and knowledge is worked with, rather than new information or knowledge being created).

While FMECA is a relevant tool for safety engineering, it does not provide a simultaneous vision of possible failures and their consequences (two failures occur at the same time on two subsystems, what is the consequence on the system as a whole?). Taking aeronautics as an example, we know that airplane crashes are rarely linked to only one failure; there are generally multiple failures occurring simultaneously.

Furthermore, FMECA as a tool must not become an end in itself. It is common for far-fetched risks to be unnecessarily associated to FMECA (i.e., somebody could break their leg skiing), or for the problems noted in the FMECA to be considered as solved problems.

FMECA has several advantages: it is a detailed, rigorous method, providing reliability prediction, automated (commercial software), relatively inexpensive. The main disadvantage is that it is not designed for hazards unrelated to failure modes (hazards related to high voltage, radiation, etc.). It is limited to external interference and influences, where little attention is given to human factors, lack of consideration of combined failures, even though, for some systems, this issue has already been addressed (20). An estimate of failure cost is often missing even though this shortage can be overcome (21).

The FMECA is a simple methodology, applicable to many facilities.

- The FMECA provides a systematic and methodical methodology.
- It needs to evaluate all possible failure modes for each component of the system.
- It is poorly suited to identify the consequences of multiple failures.
- It nonetheless highlights the specific points that should receive further study.

There are several FMECAs:

- FMECA process (identify risks of the production process).
- FMECA product (identify risks induced by the concept).
- FMECA production (identify risks linked to production facilities).
- FMECA service.
- FMECA procedure.
- FMECA sustainability.

4.7 Fault tree analysis and event tree analysis

4.7.1 Fault tree analysis

Fault tree analysis (FTA) was developed at the Bell Labs for the US Air Force in 1961 to evaluate and assess the safety for missile guidance (22). It is an analytical and deductive method (23) that derives the causes of initiating an undesired event, the top event. Used in several plants such as nuclear, chemical, and aeronautic, FTA provides both qualitative and quantitative results (24).

As input, good knowledge of design, personal training, equipment and accident history are required. FTA involves logical paths (cut-sets) and Boolean algebra to determine the causes of failure (see ▶Fig. 4.20). It is a combination of failures and their probability of occurrence applying logic gates (OR gates, AND gates). An event occurs when the gate output changes state. The output of FTA consists of the fault diagram, which exhibits the root causes of the accident at the bottom. If the fault tree is evaluated, then the frequency of the top event is also an outcome.

Some research groups have performed FTA with probabilistic risk analysis methods (25). The technique is suitable for an investigation and management of multiple cases of failures, reliability, and maintainability.

The general process to build the FTA is:

1. Set the top event (failure or an accident) to be analyzed. This event should be specific enough not to cause a too great tree, which can quickly happen.
2. Define successively, and in a top-down approach, the direct causes of each event. All causes (events entering the logic gate) necessary and sufficient for the occurrence of the consequence (event exiting the logic gate).
3. Each time a new level is built, it should be directly linked to higher levels by logic gates respecting the Boolean algebra (see ▶Fig. 4.20).
4. Finish the process of tree building when the development of events is no longer feasible or desirable.
5. Assign a likelihood of occurrence of each basic event.
6. Calculate, using Boolean algebra, likelihood of occurrence of the top event.

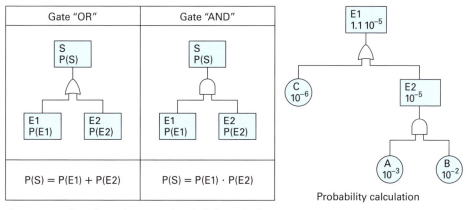

Fig. 4.20: Boolean logical gates and probability calculation.

7. Identify vulnerabilities in highlighting the critical path (cut sets), i.e. combination of the most likely basic events leading to the occurrence of the top event.
8. Propose measures that reduce the occurrence of the top event, giving priority to the most likely critical paths or addressing the common causes (basic events involved in several critical paths).

The probability calculation according to the FTA procedure is described in ▶Fig. 4.20. The steps and details of the calculation are as follows:

Known data: Probability "P" of A to happen is: $P(A)=10^{-3}$, for B it is 10^{-2} and for C is 10^{-4}.

Calculation of the probability for event E2 (logical gate AND) leads to:
$P(E2) = P(A) \times P(B) = 10^{-3} \times 10^{-2} = 10^{-5}$.

Calculation of the probability for event E1 (logical gate OR) leads to
$P(E1) = P(C) + P(E2) = 10^{-4} + 10^{-5} = 1.1 \times 10^{-4}$.

The summarized tree building is depicted in ▶Fig. 4.21, where the most important rule to remember is that the basic events of a fault tree should be strictly independent.

An example is given in ▶Fig. 4.22 for the main event "a train passes at red signal". The tree is built according to the above method, also indicating the probabilities calculations.

FTA allows considering combinations of events that can ultimately lead to a feared event, and indeed represents a deductive method. It allows a good fit with the analysis of past incidents, indicating that major accidents reported result most often from a

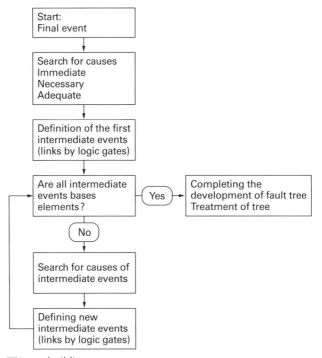

Fig. 4.21: FTA tree building.

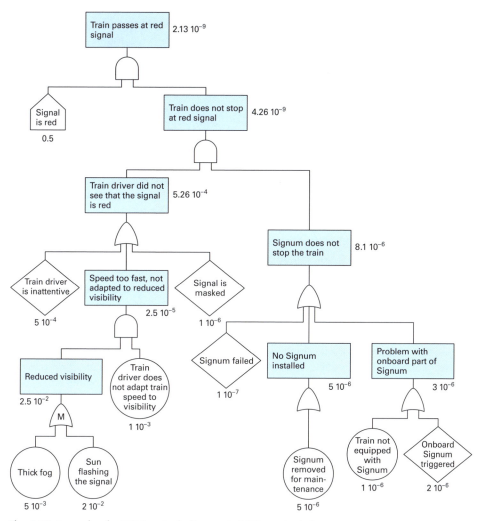

Fig. 4.22: Example of a FTA tree including probabilities calculations.

combination of several events that alone could not cause such accidents. Moreover, by estimating the event's likelihood of occurrence leading to the final event, it will provide criteria for determining priorities for the prevention of potential accidents.

The FTA approach is not suitable for the risk analysis of a complete system, including a number of adverse events that may be very large. The quantification of FTA is extremely powerful, but is often very difficult to implement because the rate of occurrence of basic events is often unknown. (All basic events without exception must be quantified in order that the top event occurrence can be calculated). Seemingly simple, this technique can easily lead the beginner to build a tree completely wrong without knowing it, delivering results and analysis that are potentially dangerously false.

> FTAs are logic block diagrams that display the state of a system (top event) in terms of the states of its components (basic events). An FTA is built top-down and in term of events rather than blocks. It uses a graphic "model" of the pathways within a system that can lead to a foreseeable, undesirable loss event (or a failure). The pathways interconnect contributory events and conditions, using standard logic symbols (AND, OR, etc.).

4.7.2 Event tree analysis

Event tree analysis (ETA) was first used in nuclear industries around 1974 to assess risks of nuclear power plants using light water. It uses an inductive approach to determine the consequences of an undesirable event via a graphical tool (4). The consequence of event follows a series of paths to which probabilities are given. The technique can be used for qualitative as well as quantitative reliability and risk analysis. The required inputs are: design knowledge, accident history, or similar scenarios. ETA processes by identifying an accident and pivotal event, building event tree diagrams and evaluating the risks. The output of an ETA can be summarized as: mishap outcome and risk probability, causal sources, and safety requirements. The main disadvantage is that it cannot study multiple failures on the same initiating event. Multiple ETAs require multiple initiating events.

The ETA method is exploratory in that it allows identifying long-term consequences that may be unknown. The methodology addresses both desirable and undesirable consequences, as it allows for the inclusion in the analysis the responses of safety systems (alarm, automatic shutdown, etc.) and operating staff. The initiating event itself may be accidental (failure, incident) or perfectly anticipated (e.g., particular operation).

The method is quantitative as it allows, provided that the likelihood of occurrence of the initiating event and failure rates of various safety systems are known, calculating the occurrence of each of the final consequences, e.g., of each scenario.

The general process to build the ETA is:

Select an initiating event.
Identify safety functions or system players that can influence the course of events caused by the initiating event.
Build the tree successively considering possible responses to the different actors.
The build process stops once all the elements which, by their reaction influencing the spread of the consequences, have been considered.
Determine the final consequence of each scenario (e.g., in terms of material, human or environmental).
Assign a probability to each branch of the tree.
Calculate by simply multiplying the likelihood of occurrence of each scenario.
Propose adapted mitigation measures if the criticality of certain scenarios is considered too high.

The different steps are:

1. Starting and final points

    - The logical development of an event tree is to answer the question "What happens if …?" This leads to a tree because usually we answer, "It depends: if … then … else …" It is this alternative that results in branches.

2. Identification of safety functions
    - Safety features must be provided as barriers in response to the initiating event. They generally aim to prevent, as far as possible, that the initiating event is the cause of a major accident.
3. Recurrence
    - Having posed an alternative, the question is repeated for each of the alternatives. It develops and branches up to get the consequences.
4. Representing
    - Traditionally, the initial event is located at the left and the tree grows from left to right. The upper branch of each alternative represents the desired operation and the lower the undesired operation.
5. Likelihood, tree exploitation
    - A quantitative aspect could be added to the event tree indicating the probabilities of events taken into account. The occurrence of each alternative is indicated in the upper branch (the other being its complement to 1).

▶Fig. 4.23 develops an ETA tree for an originating event that is the failure of a cooling system, based on the above-mentioned method.

The main limitation of the event tree is that it can only consider a single initiating event and is therefore not possible to identify scenarios for multiple initiating events. Furthermore, although more powerful than the FMECA approach to integrate the system responses, the ETA approach makes it difficult to make a systemic study because there is no systematic search of initiating events.

ETA is a method to examine, from an initiating event, the sequence of events that may or not lead to a potential accident. It is thus particularly useful for studying the architecture of safety resources (prevention, protection, response) that already exist or that may be considered on a site. It can be used for the analysis of accidents after the fact. This method can be quickly cumbersome to implement. Accordingly, one must carefully define the initiating event that will be the subject of the analysis.

> ETA is an inductive failure analysis performed to determine the consequences of a single failure for the overall system risk or reliability. It shows all possible outcomes resulting from an accidental (initiating) event, taking into account whether installed safety barriers are functioning or not, and additional events and factors.

4.7.3 Cause-consequence-analysis (CCA): a combination of FTA and ETA

CCA, often called "bowtie analysis" is a tree-like approach largely used in European countries. CCA is used in different industrial sectors by enterprises like SHELL, from where these kinds of tools originate.

As FTA is not inductive, it cannot give an overview of a system and is limited to studying specific aspects. Conversely, uninterested in the causes of failures, ETA is often

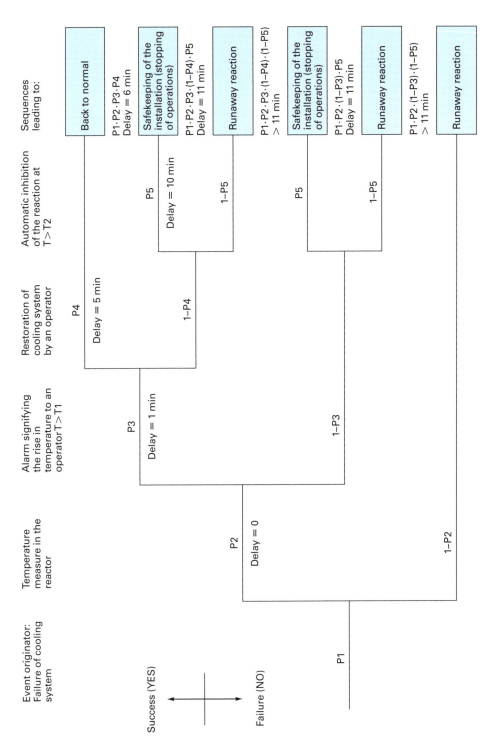

Fig. 4.23: Example of an ETA tree.

4.7 Fault tree analysis and event tree analysis

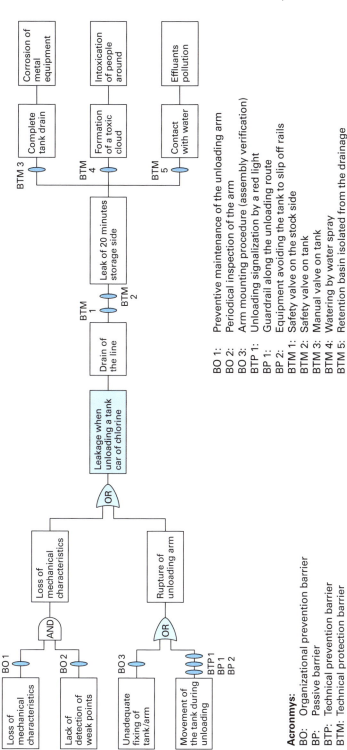

Fig. 4.24: Example of a CCA tree.

difficult to quantify as such and, although its inductive nature, the depth of analysis is often not sufficient to ensure a sufficiently accurate assessment of the level of risk.

The complementarities of these two approaches are reflected in the combined methodology FTA/ETA = CCA, which is one of the most rigorous and powerful methods. The simultaneous inductive-deductive approach allows for combining the general and the particular and has an extremely good coverage of the considered situations.

The beginning of the procedure is identical to the ETA method. One selects the desired initiating event to study and constructs the event tree qualitatively. All intermediate failure causes can be determined by constructing a fault tree that will be linked directly to a branch of the event tree.

Quantifying the structure is performed either directly at the event tree for the branches that are not bound to a fault tree, or via the quantification of fault trees for the other branches.

▶Fig. 4.24 illustrates an example of a CCA tree, starting from the main event of a leakage when unloading a tank car of chlorine. In addition, correctives measures in terms of organizational, passive, technical in prevention or protection are also depicted. The tree representation allows not only to calculate probabilities when necessary (FTA) but also to clearly indicates where are the problems and where corrective measures should apply.

In conclusion, although powerful, the implementation of a CCA analysis can be very heavy and needs to be performed by experts. This methodology is not systemic, as it suffers the same limitation of ETA analysis, namely that it can consider a single initiating event at a time. This tool clearly highlights the action of the safety barriers opposing these accident scenarios and can provide a demonstration of enhanced risk management. For a complex system comprising several tens, even hundreds, of specific failures (initiating events), it would be illusory to perform a risk analysis of the total system by the FTA/ETA approach, because it will represent years of work. It provides a concrete visualization of accident scenarios that could occur starting from the initial causes of the accident to the consequences at the level of identified targets.

> Cause-consequence analysis is a technique that combines the ability of fault trees to show how various factors may combine to cause a hazardous event with the ability of event tress to show the various possible outcomes. Sequences and therefore time delays can be illustrated in the consequence part of the diagram. A symbolism similar to that in fault trees is used to show logical combinations. The technique has considerable potential for illustrating the relationships from initiating events through to end outcomes. It can be used fairly directly for quantification, but the diagrams can become extremely unwieldy.

4.8 The risk matrix

The "risk assessment decision matrix", often shortened to "the risk matrix", is a systematic approach for estimating and evaluating risks. This tool can be employed to measure and categorize risks on an informed judgment basis as to both probability and consequence and as to relative importance. An example of the risk matrix is shown in ▶Tab. 4.15.

Tab. 4.15: The risk assessment decision matrix.

Severity of consequences	Likelihood of risk					
	F	E	D	C	B	A
	Impossible	Improbable	Remote	Occasional	Probable	Frequent
I Catastrophic						
II Critical						
III Marginal						
IV Negligible						

Source: based on Department of Defense, 2000 (26).

Once the risks have been identified, the question of assigning consequence and likelihood ratings must be addressed. A common, basic example of assigned ratings by a team on a generalized basis can be found in ▶Tab. 4.16.

Tab. 4.16: Criticality and frequency rating for the risk assessment decision matrix.

Severity of consequences – ratings		
Category	Descriptive word	Results in either
I	Catastrophe	• An on-site or an off-site death • Damage and production loss greater than €750,000
II	Critical	• Multiple injuries • Damage and production loss between €75.000 and €750,000
III	Marginal	• A single injury • Damage and production loss between • €7,500 and €75,000
IV	Negligible	• No injuries • Damage and production loss less than €7,500

Hazard probability – ratings		
Level	Descriptive word	Definition
A.	Frequent	Occurs more than once per year
B	Probable	Occurs between 1 and 10 years
C	Occasional	Occurs between 10 and 100 years
D	Remote	Occurs between 100 and 10,000 years
E	Improbable	Occurs less often than once per 10,000 years
F	Impossible	Physically impossible to occur

Source: based on Department of Defense, 2000 (26).

Tab. 4.17: Definitions and recommended actions for rankings.

Ranking	Description	Required action
1	Unacceptable	Should be mitigated with technical measures or management procedures to a risk ranking of three or less within a specified time period such as 6 months.
2	Undesirable	Should be mitigated with technical measures or management procedures to a risk ranking of three or less within a specified time period such as 12 months.
3	Acceptable with controls	Should be verified that procedures or measures are in place.
4	Acceptable	No mitigation action required.

The probability level F, "impossible", makes it possible to assess residual risks for cases in which the hazard is designed out of the system.

The rankings provide a quick and simple priority sorting method. The rankings are then given definitions that include, e.g., definitions and recommended actions similar to those in ▶Tab. 4.17.

Such a safety risk assessment methodology is especially implemented in case of low-likelihood, high-consequence risks.

It is very important that likelihood estimates as well as consequence estimates are very well-considered and carried out by experienced risk managers. When this has been done, a risk is "binned" into one of the squares of the risk matrix as a function of the level of its consequence and the level of its likelihood (occurrence probability). How then can risks then be ranked or prioritized across the squares of the risk matrix? To answer this question, consider e.g., the 4 × 4 ordinal risk matrix of ▶Fig. 4.25.

Denote an *(i, j)* risk event as one that has level *i* consequence and a level *j* likelihood, where *i, j* = 1, 2, 3, 4. A common approach to prioritizing risks is to multiply the consequence and likelihood levels that define each square and use the resultant product to define the square's score. ▶Fig. 4.26 displays a matrix with risk resultant scores.

Risks binned into a specific square then receive that square's score and are ranked accordingly. Risks with higher scores have higher priority than risks with lower scores. However, there are some disadvantages to this approach.

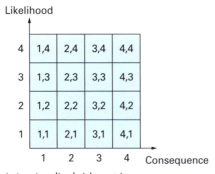

Fig. 4.25: A 4 × 4 ordinal risk matrix.

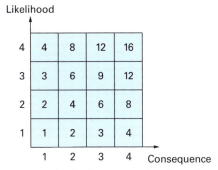

Fig. 4.26: A 4 × 4 ordinal risk matrix with risk resultant scores based on multiplication.

The first disadvantage that would possibly lead to problems is the multiplication of ordinal numbers, which is not a permissible arithmetic operation. The second disadvantage is that a risk with a level 4 impact and a level 1 likelihood receives the same score as a risk with a level 1 impact and a level 4 likelihood. Although these are two very different risks, they are equally valued and tie in their scores if the multiplication method is used. In industrial practice, risk managers should not lose visibility in the presence of high-consequence risks (type II and III risks) regardless of their likelihood. The third disadvantage is that there are numerous ties in the scores of ▶Fig. 4.26, and, in view of the problem just discussed, the question should be posed whether risks that tie in their scores, should be equally valued, especially when these ties occur in very different consequence-likelihood regions of the risk matrix. Hence, how to solve these problems?

As Garvey (27) indicates, one way is to first examine the risk attitude of the risk management team. Is the team consequence-averse or likelihood-averse? A strictly impact-averse team is not willing to trade-off consequence for likelihood. For such a team, low-likelihood high-consequence risks should be ranked high enough so that they remain visible to management. But how can risk prioritizations be assigned in this regard? The following presents an approach to address these considerations.

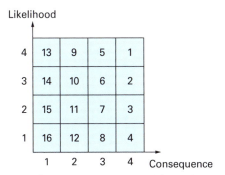

Fig. 4.27: A strictly consequence-averse risk matrix.

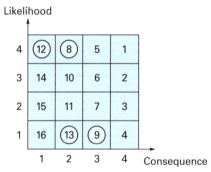

Fig. 4.28: First iteration of risk ranking.

To start, the ordinal risk matrix should be redefined in a way that prioritizes risks along a strictly consequence-averse track. ▶Fig. 4.27 shows a matrix where this has been carried out.

Each square is scored in order of consequence, criticality by the risk management team. The lower the score, the higher the priority. Risks categorized in the right-upper corner of the matrix have the highest priority and have a score of 1. Note that in this risk matrix, risks with a level 4 consequence will always have higher priority than those with a level 3, and so forth. Thus, this matrix is one where risks with the highest consequence will always fall into one of the first four squares – and into a specific square as a function of their judged likelihood.

Now, a risk management team might want to deviate from this strictly consequence-averse risk matrix [see also (27)]. To illustrate this, imagine that the team decided that any risk with a level 4 consequence should remain ranked in one of the first four squares of the risk matrix (i.e., the right-most column). For all other columns, trade-offs could be made between the bottom square of a right-hand column, and the top square of its adjacent left column (first iteration). This is shown by the circled squares in the risk matrix illustrated in ▶Fig. 4.28.

Suppose that the team decides to further refine the rank-order of the squares in the matrix. This second iteration is illustrated by circles in ▶Fig. 4.29.

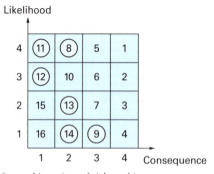

Fig. 4.29: Second iteration of risk ranking.

Tab. 4.18: The risk assessment decision matrix.

Severity of consequences	Probability of Hazard					
	F Impossible	E Improbable	D Remote	C Occasional	B Probable	A Frequent
I Catastrophic						1.
II Critical					2.	
III Marginal				3.		
IV Negligible			4.			
Risk code/ actions	1. Un-acceptable	2.	Un-desirable 3.	Acceptable with controls	4. Acceptable	

This discussion illustrates one approach for directly ranking or prioritizing risks on the basis of their consequences and likelihood.

Finally, it is common practice to assign color bands within a risk matrix. These bands are intended to reflect priority groups. ▶Tab. 4.16 uses the matrix from ▶Tab. 4.15 to illustrate how the matrix from ▶Fig. 4.29 might be colored with respect to priority groups. Thus ▶Tab. 4.18 is created.

> Risk ranking uses a matrix that has ranges of consequence and likelihood as the axes. The combination of a consequence and likelihood range gives an estimate of risk or a risk ranking. Without adequate consideration of risk tolerability, a risk matrix can be developed that implies a level of risk tolerability much higher than the organization actually desires. Risk tolerability should thus be carefully considered by the risk management team.

4.9 Quantitative risk assessment (QRA)

Quantitative risk assessment (QRA) is a very systematic approach to calculate individual fatality risks and societal risks from industrial facilities or parts thereof. The generic stages for the QRA procedure for the risk assessment of hazards are:

- System description – context and scope
- Hazard/threat identification
- Incident enumeration
- Selection; worst-case and/or credible accident scenario identification
- Consequence assessment
- Likelihood assessment
- Risk assessment
- Visualization and utilization of risk estimates

In the stage 1, the scope of the study is established. Information regarding site locations, environs, weather data, process flow diagrams, layout drawings, operating and maintenance procedures, technology documentations, etc. is gathered, as well as data regarding (for example) the use, storage, processing, etc. of hazardous materials. Note that not only information from within the study scope area, but also from outside the scope area needs to be collected. This information is then further employed during the different consecutive steps of the QRA. Stage 2 identifies all possible hazards and threats that could possibly lead to incidents and accidents. There are many possible hazard identification techniques available, such as experience, engineering codes, checklists, detailed knowledge of the situation/plant/process/installation, equipment failure experience, hazard index techniques, what-if analysis, HAZOP studies, FMEA, preliminary hazard analysis, etc. (see also previous sections). Stage 3 of a QRA procedure comprises the identification and tabulation of all incidents, without regard to importance or initiating event. Based on the list of incidents from the previous stage, stage 4 selects one or more significant incidents to represent all identified incidents. Mostly, the incidents chosen are most credible or worst-case. Hence, the credible and/or worst-case accident scenarios are selected in this stage. In industrial practice, often a large number of scenarios (sometimes up to 1,500, depending on the scope of the study) still remain, out of the very large number of possible incidents. Having established the scenarios that need further processing, stage 5 first determines the potential for damage or injury from specific events. CCPS (28) indicates that a single accident (e.g., rupture of a pressurized flammable liquid tank) can have many distinct accident outcomes (e.g., vapor cloud explosion, BLEVE, flash fire). These possible outcomes can then be analyzed using source and dispersion models, explosion and fire models, and other effect models, where the consequences to people and infrastructure is determined. In stage 6, a methodology is used to estimate the frequency or probability of occurrence of an accident. Estimates may, e.g., be obtained from historical incident and accident data on failure frequencies, or from failure sequence models such as fault trees and event trees. Most systems require consideration of factors such as common-cause failures, human reliability, external events, etc. Stage 7 combines the results obtained in the previous two stages, i.e., the consequences and the likelihood of all accident outcomes from all selected scenarios, and uses these to obtain a risk estimate for every scenario. The risks of all selected accident scenarios are consecutively summed up to provide an overall measure of risk. In the last stage the results are tabulated and visualized using, e.g., iso-risk contours (for individual fatality risks) or FN-curves (for societal risks), to be able to help decision-makers. Decisions are made either through relative ranking of risk reduction strategies or through comparison with specific risk targets. The last stage obviously requires some risk acceptability criteria, or risk guidelines, to be elaborated in advance by the user or the authorities.

Obviously, a QRA study is not a "simple" study, and the stepwise plan explained above indicates that the aim of the study is to narrow down an initial abundance of possible scenarios to an intelligently considered and justifiable number of accident scenarios, calculate them through, and obtain a (relatively) limited number of results where one is able to make objective choices. There are some disadvantages attached to this approach. The main critiques of QRA is that it is unrealistic: it often deals with absurdly small numbers and statistics, which can often lead observers to question the

4.9 Quantitative risk assessment (QRA)

validity of the approach. QRA tries to find out what circumstances have to happen simultaneously to lead to a serious problem, assesses which of the circumstances has the greatest importance in the hazard, and suggesting that this would be the primary focus of risk management. Another critique is that it is not reproducible. This is because a QRA relies on the application of generic data where no specific data is available, and largely depends on choices of the user. The many different choices that have to be made (regarding the consequence models, scenario selection, frequency estimates, etc.) may lead to different results, even for an identical scope/situation. This is an important drawback as the technique is developed to provide consistent results. Compared to other techniques, the results will be more consistent, but, as mentioned, they should still be interpreted with caution. Furthermore, there are arguments that the results of a QRA are best used to compare the relative safety of different systems and not look at the absolute magnitude of the risk in relation to risk criteria.

Despite all the shortcomings of a QRA, and despite the obvious fact that it is not an exact description of reality, it can be the best available tool to date to assess the risks of complex facilities and situations where a large number of accident outcome cases are at hand, and at the same time sophisticated models are required.

Using the cube depicted in ▶Fig. 4.30, the QRA risk analysis technique can be positioned in relation to other methodologies. The cube is based on three essential parameters:

- Focus
- Complexity
- Number of incident scenarios investigated

An example of a QRA study is difficult to provide, as the technique is characterized by a very elaborated study for several steps. Incident enumeration, scenario selection,

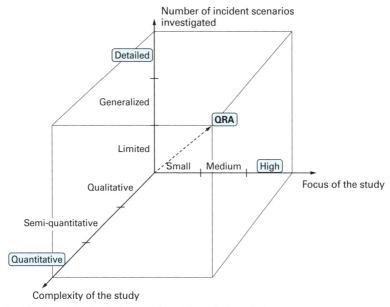

Fig. 4.30: Positioning of QRA relative to other risk analysis techniques.

consequence assessment/calculation, and likelihood assessment/determination can all be very complicated and time-consuming studies. Moreover, different software packages can be chosen and used to help the user carry out several required calculations of a QRA.

> QRA is a formalized specialist method used for calculating numerical individual and societal (employee and public) risk level values for comparing risks, or for comparison with regulatory risk criteria (e.g., in case of land-use planning). To do a QRA, the value of the potential losses needs to be determined. Then the probability of the occurrence of the risk failure needs to be estimated. Finally, the (annual) loss expectancy is calculated.

4.10 Layer of protection analysis

Layer of protection analysis (LOPA) is an analytical procedure that looks at the safeguards to see if the protection provided is adequate for every known risk. It shows whether additional controls or shutdown system(s) are required, by comparing the risks with and without these additional elements, against a set of predetermined criteria (29). LOPA basically answers three questions:

1. How safe is safe enough?
2. How many layers of protection are needed?
3. How much risk reduction does every layer of protection realize?

If a LOPA is carried out effectively, an accident scenario screened and assessed by the technique should not be able to lead to an accident.

The objective of LOPA is thus to determine if there are sufficient layers of protection against a certain accident scenario, and thus to answer the question whether a risk can be accepted or tolerated. Despite the fact that often many layers of protection exist to prevent accident scenarios, no layer is perfectly effective (e.g., the Swiss cheese model by Reason and the holes in the cheese), and therefore accidents still may occur.

By using order of magnitude categories for initiating event frequency, consequence severity, and the likelihood of failure of independent protection layers, the risk of a well-defined accident scenario is determined. Hence, LOPA is not a fully quantitative risk assessment approach, but is rather a simplified method for assessing the value of protection layers for a certain accident scenario; by assessing the risks dependent on various protection layers, LOPA assists in deciding between alternative choices for risk mitigation (30). The typical layers of protection that can be used in LOPA are depicted in Chapter 3.

An accident scenario is typically identified during a qualitative hazard evaluation, management of change evaluation, or design review. LOPA is limited to evaluating single cause-consequence pairs as a scenario, and provides an order of magnitude approximation of the risk of any scenario. Once a scenario is determined, the risk analyst can use LOPA to establish which controls (often called "safeguards") meet the definition of IPLs, and then estimate the scenario order of magnitude risk. In LOPA, the scenario that is chosen is usually the worst-case one.

4.10 Layer of protection analysis

In many LOPA applications, the risk manager has to identify all scenarios: cause-consequence pairs, exceeding a certain predetermined tolerance of risk. In other applications, the risk manager chooses the scenario that likely represents the highest risk from many similar scenarios. As already mentioned, LOPA typically starts with a list of possible scenarios determined by qualitative risk analyses. The further stages of the technique are:

- Identify the consequence to screen the scenario.
- Select an accident scenario.
- Identify the initiating event of the scenario and establish the initiating event's frequency (events per year).
- Identify the IPLs and estimate the probability of failure on demand of each IPL.
- Estimate the risk of a scenario by mathematically combining the consequence, initiating event, and IPL data.

In stage 1, the outcome/impact/magnitude of the consequence of a scenario (provided by, e.g., a HAZOP study) is estimated and evaluated. Different possible outcomes can be taken into account from an initiating event (e.g., a release): e.g., the impact to people, the environment, and production. Stage 2 is concerned with choosing one scenario that can be represented by a single cause-consequence pair. Each scenario consists of the following elements: (i) an initiating event that initiates the chain of events; (ii) a consequence that results if the chain of events continues without interruption; (iii) enabling events or conditions that have to occur or be present before the initiating event can result in a consequence; (iv) the failure of safeguards (note that not all safeguards are IPLs, but all IPLs are safeguards!) Scenarios may also include probabilities of a different kind (ignition probability of flammable materials, probability that a fatal injury will result from exposure of effects of fire, explosion, etc.). In stage 3, a frequency has to be estimated where information and casuistic may be insufficient. Therefore, companies and sometimes authorities provide guidance on estimating the frequency to achieve consistency in LOPA results. Stage 4 forms the heart of the LOPA method. Here, the existing safeguards that meet the requirements of IPLs for a given scenario need to be recognized. Some accident scenarios will require only one IPL, while other accident scenarios may require many IPLs, or IPLs of very low or high PFDs, to achieve a tolerable risk for the scenario. In the final and fifth stage, the scenario risk is calculated. This may be carried out by different approaches. Regardless of the formulae and graphical methods available to do so, companies usually have a standard form for documenting the results.

▶Fig. 4.31 illustrates the use of IPLs in a LOPA scenario. One of the main characteristics of any IPL is its reliability, which is expressed by its PFD. As indicated in Chapter 3, the PFD is dimensionless, and may vary between 10^{-1} (weak protection) to 10^{-5} (very strong protection). Mostly, a value of 10^{-3} is considered as a strong protection. A value of PFD equal to 1 indicates that the layer of protection does not contribute to the LOPA scenario.

In the illustrative example of ▶Fig. 4.31 (where every IPL is characterized by a PFD equal to 0.1), the probability that the initiating event eventually leads to the cause-consequence pair envisioned, that is, the probability that the scenario takes place, is estimated at 0.0001 or 10^{-4}.

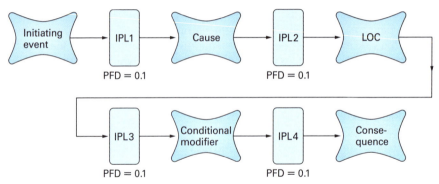

Fig. 4.31: Use of IPLs in a LOPA scenario.

In summary, a LOPA analysis is aimed at assessing the efficiency (or the lack thereof) of safety and protection measures to prevent accident scenarios. It allows for comparing a variety of technical, organizational, and other protective measures. The measure leading to the highest risk reduction level can be determined for any single cause-consequence pair where sufficient information is available or retrievable.

> LOPA is an analytical procedure that looks at the safeguards to see if the protection provided is adequate for every known risk. It is a powerful analytical tool for assessing the adequacy of protection layers used to mitigate process risk. LOPA builds upon well-known process hazards analysis techniques, applying semi-quantitative measures to the evaluation of the frequency of potential incidents and the probability of failure of the protection layers.

4.11 Bayesian networks

A Bayesian network (BN) is a network composed of nodes and arcs, where the nodes represent variables and the arcs represent causal or influential relationships between the variables. Hence, a BN can be viewed as an approach used to have an overview of relationships between causes and effects of a system. In addition, each node/variable has an associated conditional probability table (CPT). Thus, BNs can be defined as directed acyclic graphs, in which the nodes represent variables, arcs signify direct causal relationships between the linked nodes, and the CPTs assigned to the nodes specify how strongly the linked nodes influence each other (31). The key feature of BNs is that they enable us to model and reason about uncertainty (32).

The nodes without any arc directed into them are called "root nodes", and they are characterized with marginal (or unconditional) prior probabilities. All other nodes are intermediate nodes and each one is assigned a CPT. Among intermediate nodes, the nodes with arcs directed into them are called "child nodes" and the nodes with arcs directed from them are called "parent nodes". Each child has an associated CPT, given all combinations of the states of its parent nodes. Nodes without any child are called "leaf nodes".

4.11 Bayesian networks | 141

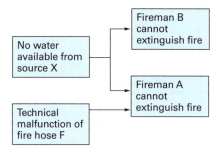

Fig. 4.32: Illustrative example of a Bayesian network.

As an illustrative example and to keep it as simple as possible, we assume, in ▶Fig. 4.32, that the variables are discrete, hence, they are either "true" or "false". Of course, in industrial practice, some variables such as "No water available from source X" could be continuous on a scale from zero to a certain possible maximum flow of water.

▶Fig. 4.32 illustrates that a technical failure of the fire hose F can cause Fireman A not to be able to extinguish the fire. Also, water not being available from source X may cause (or influence) both Fireman A and Fireman B to not be able to extinguish the fire. In this example, no water available from source X does not imply that Fireman B will definitely not be able to extinguish the fire, as he may tap water from a different source – e.g., source Y – but there is an increased probability that he will not be able to extinguish the fire. The CPT for the node/variable "Fireman B cannot extinguish fire" holds this information, providing the conditional probability of each possible outcome given each combination of outcomes for its parent nodes. ▶Fig. 4.33 shows a CPT for "Fireman B cannot extinguish fire".

▶Fig. 4.33 illustrates that, e.g., the probability that Fireman B cannot extinguish the fire, given that there is water available from source X, is 0.1. Obviously, the CPTs for the root nodes are very simple (they have no parents), and we only have to assign a probability to the two states "true" and "false" (see ▶Fig. 4.34). On the contrary, the CPT for Fireman A is more complicated as this node has two parents, leading to four combinations of parent states (see ▶Fig. 4.35).

The probabilities that have to be assigned to the different states of the variables can be determined in different ways. One way is to use historic data and statistical information (observed frequencies for example) to calculate the probabilities. Another way,

		No water available from source X	True	False
Fireman B cannot extinguish fire	True		0.8	0.1
	False		0.2	0.9

Fig. 4.33: CPT for "Fireman B cannot extinguish fire".

Technical malfunction of fire hose F	True	0.4
	False	0.6

No water available from source X	True	0.1
	False	0.9

Fig. 4.34: CPTs for "Technical malfunction of fire hose F" and "No water available from source X".

		Technical malfunction of fire hose F	True		False	
		No water available from source X	True	False	True	False
Fireman A cannot extinguish fire	True		0.8	0.6	0.6	0.3
	False		0.2	0.4	0.4	0.7

Fig. 4.35: CPT for "Fireman A cannot extinguish fire".

especially when no or insufficient statistical information is available, is to use expert opinion. Hence, both probabilities based on objective data and subjective probabilities can be used in BN.

BN are based on Bayes' Theorem and Bayes theory to update initial beliefs or prior probabilities of events using data observed from the event studied. Bayes' Theorem can be expressed as follows:

$$P(\text{Event, info}) = \frac{P(\text{Event}) \cdot P(\text{info} | \text{Event})}{P(\text{info})}$$

In this equation, $P(\text{Event})$ is the prior probability of an event, $P(\text{Info} | \text{Event})$ is the likelihood function of the event, $P(\text{info})$ is the probability of info/data observed (commonly called as evidence), and $P(\text{Event} | \text{Info})$ is the posterior probability of the event. Hence, we start with a prior probability of the event, but we are interested in knowing what is the posterior probability of the event, given the evidence "info". This can be done using Bayes' Theorem.

Let us apply this theorem to our example. Assume that we find out that Fireman B is indeed not able to extinguish the fire. Then, intuitively, we feel that the probability of water not being available from source X, must have increased from its prior value of 0.1 (see ▶Tab. 4.34). But by how much? Bayes' Theorem provides the answer. We know e.g., from the CPT that P(Fireman B cannot extinguish fire | no water available for source X) = 0.8 and that P(no water available from source X) = 0.1. So the numerator in Bayes' Theorem is 0.08. The denominator, P(Fireman B cannot extinguish fire), is the marginal (or "unconditional", as already mentioned) probability that Fireman B is not able to extinguish the fire: it is the probability that Fireman B cannot extinguish the fire when we do not know any specific information about all variables or events that influence it (in the case of our example, the availability of water from source X). Although this value cannot be directly determined, it can be calculated using probability theory, and equals 0.17. Hence, substituting this value in Bayes' Theorem, we obtain:

P(no water available for source X | Fireman B cannot extinguish fire) = 0.08/0.17 = 0.471.

Therefore, the observation of the Fireman B not being able to extinguish the fire, significantly increases the probability that there is no water available from source X: from 0.1 to 0.471. In a similar way, we are able to use the information to calculate the revised belief of Fireman A not being able to extinguish the fire. Once we know that Fireman B is not able to extinguish the fire, the prior probability of 0.446 (for Fireman A not being able to extinguish the fire) increases to 0.542. In fact, the calculations (which can be quite cumbersome), are done automatically in any BN tool.

Updating the probabilities with evidence and new information is called "propagation". Any number of observations can be input anywhere in a Bayesian network, and propagation can be employed to update the marginal probabilities of all the unobserved variables.

As Fenton and Neil (32) explain, Bayesian networks offer several important benefits when compared with other available techniques. In BN, causal factors are explicitly modeled. In contrast, in regression models historical data alone are used to produce equations relating dependent and independent variables. No expert judgment is used when insufficient information is available, and no causal explaining is carried out. Similarly, regression models cannot accommodate the impact of future changes. In short, classical statistics (e.g., regression models) are often good for describing the past, but poor for predicting the future. A BN will update the probability distributions for every unknown variable whenever an observation or evidence is entered into any node. Such technique of revised probability distributions for the cause nodes as well as for the effect nodes is not possible in any other approach. Moreover, predictions are made with incomplete data. If no observation is entered then the model simply assumes the prior distribution. Another advantage is that all types of evidence can be used: objective data as well as subjective beliefs.

This range of benefits, together with the explicit quantification of uncertainty and ability to communicate arguments easily and effectively, makes BNs a powerful solution for all types of risk assessment.

Whereas the essential BN is static, in many real-world problems, the changes in values of uncertain variables need to be modeled over successive time intervals. Dynamic Bayesian Networks (DBN) extends static BN algorithms to support such modeling and inference. Hence, temporal behavior can be captured by DBNs, and the necessary propagation is carried out on compact (rather than expanded, unmanageable and computationally inefficient static) models.

> The scenario in a risk analysis can be defined as the propagating feature of a specific initiating event which can go to a wide range of undesirable consequences. If we take various scenarios into consideration, the risk analysis becomes more complex than it does without them. Bayesian networks allow for considering the effects of predictions by incorporating knowledge with data to develop and use causal models of risk that provide powerful insights and better decision-making.

4.12 Conclusion

The field of risk analysis has assumed increasing importance in recent years given the concern by both the public and private sectors in safety, health, security and environmental problems. The approaches to risk assessment normally center on identifying a risk, and then assessing the extent to which the risk would be a problem (the severity of the risk), and secondly, by calculating how likely it is that the risk would occur (the probability). Following this, attempts are made to mitigate the risk (ideally eliminate the source or take measures to reduce the likelihood of it happening). Where the possibility of the risk occurring can be reduced no further, steps should be taken to monitor the risk (detection), preferably as an early warning, so that action can be taken.

Several risk analysis techniques were presented in this chapter. Qualitative methodologies, though lacking the ability to account the dependencies between events, are effective in identifying potential hazards and failures within the system. The tree-based techniques addressed this deficiency by taking into consideration the dependencies between each event. The probabilities of occurrence of the undesired event can also be quantified with the availability of operational data. Methods such as QRAs and Bayesian Networks use quantitative data to make as accurate predictions as possible.

There is no universal risk analysis method. Each situation analyzed will require the adequate method, being only descriptive, qualitative, semi-qualitative, semi-quantitative, quantitative or predictive, as well as the static and dynamic one. Successful risk analyses require scientists and engineers to undertake assessments to characterize the nature and uncertainties surrounding a particular risk. It is up to the team performing the risk analysis to use the correct tool, in a similar way as a carpenter would not use a screwdriver to hammer a nail even if he will succeed eventually.

Regardless of the prevention techniques employed, possible threats that could arise inside or outside the organization need to be assessed. Although the exact nature of potential disasters or their resulting consequences are difficult to determine, it is beneficial to perform a comprehensive risk assessment of all threats that can realistically occur at – and to – the organization.

We underlined that in undertaking risk assessments, it is important to attempt risk mitigation and to attempt to lower the risk until the risk can be lowered no further. This involves identifying actions to reduce the probability of an event and to reduce its severity. When this can be taken no further, the focus should move towards providing a more reliable detection method designed to initiate a reliable response to a risk event. A further important consideration is that risk-assuming actions should be periodically reassessed.

References

1. Reniers, G.L.L., Dullaert, W., Ale, B.J.M., Soudan, K. (2005) Developing an external domino accident prevention framework: Hazwim. J. Loss Prev. Proc. 18: 127–138.
2. Groso, A., Ouedraogo, A., Meyer, Th. (2011) Risk analysis in research environment. J. Risk Res. 15:187–208.
3. Ericson, C.A. (2005) Hazard Analysis Techniques for System Safety. Fredericksburg, VA: J. Wiley & Sons.
4. Ahmadi, A., Soderholm, P. (2008) Assessment of operational consequences of aircraft failures: Using event tree analysis. In: IEEE Aerospace Conference, vols. 1–9, 3824–3837.
5. CCPS (2008) Guidelines for Hazard Evaluation Procedures. New York: J. Wiley & Sons.
6. Swann, C.D., Preston, M.L. (1995) Twenty-five years of HAZOPs. J. Loss Prevent. Process Industries 8:349–353.
7. Warner, F. (1975) Flixborough disaster. Chem. Eng. Prog. 71:77–84.
8. Cagno, E., Caron, F., Mancini, M. (2002) Risk analysis in plant commissioning: The multilevel HAZOP. Reliab. Eng. Syst. Safety 77:309–323.
9. Cocchiara, M., Bartolozzi, V., Picciotto A., Galluzzo, M. (2001) Integration of interlock system analysis with automated HAZOP analysis. Reliab. Eng. Syst. Safety 74:99–105.
10. Labovsky, J., Svandova, Z., Markos, J., Jelemensky, L. (2008) HAZOP study of a fixed bed reactor for MTBE synthesis using a dynamic approach. Chem. Pap. 62:51–57.
11. Mushtaq, F., Chung, P.W.H. (2000) A systematic HAZOP procedure for batch processes and its application to pipeless plants. J. Loss Prev. Proc. 13:41–48.
12. Ruiz, D., Benqlilou, C., Nougues, J.M., Puigjaner, L., Ruiz, C. (2002) Proposal to speed up the implementation of an abnormal situation management in the chemical process industry. Ind. Eng. Chem. Res. 41:817–824.
13. McCoy, S.A., Zhou, D.F., Chung, P.W.H. (2006) State-based modeling in hazard identification. Appl. Intell. 24:263–279.
14. Zhao, C., Bhushan, M., Venkatasubramanian, V. (2005) Phasuite: An automated HAZOP analysis tool for chemical processes. Part I: Knowledge engineering framework. Process Saf. Environ. Prot. 83:509–532.
15. Chin, K.S., Chan, A., Yang, J.B. (2008) Development of a Fuzzy FMEA based product design system. Int. J. Adv. Manuf. Tech. 36:633–649.
16. Franceschini, F., Galetto, M. (2001) A new approach for evaluation of risk priorities of failure modes in FMEA. Int. J. Prod. Res. 39:2991–3002.
17. Scipioni, A., Saccarola, G., Centazzo, A., Arena, F. (2002) FMEA methodology design, implementation and integration with HACCP system in a food company. Food Control 13:495–501.
18. Thivel, P.X., Bultel, Y., Delpech, F. (2008) Risk analysis of a biomass combustion process using MOSAR and FMEA methods. J. Haz. Mat.151:221–231.
19. Su, C.T., Chou, C.J. (2008) A systematic methodology for the creation of Six Sigma projects: A case study of semiconductor foundry. Expert. Syst. Appl. 34:2693–2703.
20. Price, C.J., Taylor, N.S. (2002) Automated multiple failure FMEA. Reliab. Eng. Syst. Safety 76:1–10.
21. D'Urso, G., Stancheris, D., Valsecchi, N., Maccarini, G., Bugini, A. (2005) A new FMEA approach based on availability and costs. In: Advanced Manufacturing Systems and Technology. Kuljanic, E., editor. New York: Springer-Verlag.
22. Ericson, C.A. (1999) Fault tree analysis–a history. A history from the Proceedings of the 17th International System Safety Conference. Orlando, Florida, USA.
23. Hauptmanns, U., Marx, M., Knetsch, T. (2005) GAP – a fault-tree based methodology for analyzing occupational hazards. J. Loss Prev. Proc. 18:107–113.

24. Khan, F.I., Abbasi, S.A. (1998) Techniques and methodologies for risk analysis in chemical process industries Process Safety Prog. 11:261–277.
25. Ekaette, E., Lee, R.C., Cooke, D.L., Iftody, S., Craighead, P. (2007) Probabilistic fault tree analysis of a radiation treatment system. Risk Analysis 27:1395–1410.
26. Department of Defense (2000) Standard Practice for System Safety. MIL-STD-882D. Wright-Patterson Air Force Base, OH: HQ Air Force Material Command.
27. Garvey, P.R. (2009) Analytical Methods for Risk Management. A Systems Engineering Perspective. Boca Raton, FL: CRC Press.
28. Center for Chemical Process Safety (2000) Guidelines for Chemical Process Quantitative Risk Analysis. 2nd edn. New York: American Institute of Chemical Engineers.
29. Crawley, F., Tyler, B. (2003) Hazard Identification Methods, European Process Safety Centre. Rugby, UK: Institution of Chemical Engineers.
30. Center for Chemical Process Safety (2001) Layer of Protection Analysis. Simplified Process Risk Management. New York: American Institute of Chemical Engineers.
31. Torres-Toledano, J.G., Sucar, L.E. (1998) Bayesian networks for reliability analysis of complex systems. Lect. Notes Comput. Sci. 1484:195–206.
32. Fenton, N., Neil, M. (2007) Managing Risk in the Modern World. Applications of Bayesian Networks. London: London Mathematical Society.

5 Risk treatment/reduction

5.1 Introduction

Risk reduction encompasses all strategies whose goal is to limit the risks and damages related to a specific domain. Risk reduction is the promotion of health and assets. It is empowerment, and to quote a Canadian concept: "its role is to give the community the means to assume its destiny" (1).

Techniques of control and risk reduction aim to minimize the likelihood of occurrence of an unwanted event (taking into account the exposure to risks leading to the event) and the severity of potential losses as a result of the event; or aim to make the likelihood or outcome more predictable. They could be listed as:

- Substitution: by replacing substances and procedures with less hazardous ones, by improving construction work, etc.
- Elimination of risk exposure: this consists of not creating or of completely eliminating the condition that could give rise to the exposure.
- Prevention: combines techniques to reduce the likelihood/frequency of potential losses. Observation and analysis of past accidental events enable the improvement and intensification of prevention measures.
- Reduction/mitigation: are techniques whose goal is to reduce the severity of accidental losses when an accident occurs.
- Measures applied before the occurrence of the event (often also have an effect on the likelihood/frequency).
- Measures applied after the occurrence of the event (often aim to accelerate and enhance the effectiveness of the rescue).
- Segregation summarizes the techniques that are to minimize the overlapping of losses from a single event. It may imply very high costs.
  - Segregation by separation of high-risk units.
  - Segregation by duplication of high-risk units.

Transfer, risk transfer by:

- Contractual transfer of the risk financing, essentially insurance.
- Risk financing by retention (auto financing), finance planning of potential losses by your own resources.
- Alternative risk. The alternative risk transfer (ART) solutions comprise both elements of auto financing and contractual transfer and so cannot be classified in any of the above categories.

The simplified model of events describes in a concise way the non-desired event. It is composed of distinct parts, of which the succession leads to the event as illustrated in ▶Fig. 5.1.

To physically model an accident pathway, we must have a sequence starting from a potential energy, and a process using this potential energy. Then an incident may happen

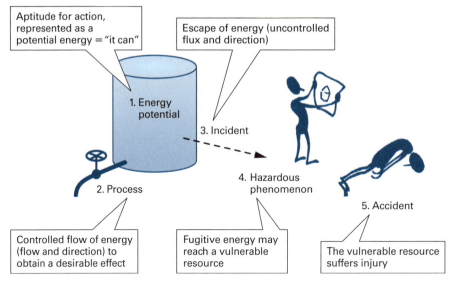

Fig. 5.1: Physical modeling of unwanted events.

as an escape of this energy in an uncontrolled flux and direction. If this escape of energy reaches a vulnerable resource (hazardous phenomenon), an accident happens. At the end, the vulnerable resource suffers from damage or injury.
Potential energy can be:

- Physical: mechanical, electrical, thermal, radiations (ionizing or not), pressure, etc.
- Chemical: reactivity, activity, corrosivity, flammability, etc.
- Physicochemical: phase change (volatility), division state (granulometry), etc.
- Physiological: sensitization, irritation, skin penetration, toxicity, etc.
- Biological: potential pathogen, mutagen, etc.

Risks are estimated according to two criteria: the probability of the unwanted event and the severity of the consequences of the unwanted event. Employers have the responsibility to take adequate prevention and\or protection and mitigation measures to avoid these accidents and\or to decrease their gravity, respectively.

Different safety measures are used to mitigate risks (2):

- Disincentives for avoiding the implementation of potential threats.
- Preventive measures inhibiting a threat realization (they act on causes and reduce the probability of occurrence).
- Protective (and mitigation) measures limiting the impact of a threat by reducing the direct consequences (they act on consequences without taking into account the occurrence).
- Remedial actions limiting the consequences of the threat implementation and indirect consequences.
- Recovery measures aiming to recover some of the damage by transferring the risk (e.g., insurance) to limit final losses.

5.1 Introduction

Risk reduction aims to influence the occurrence of a hazard and/or its consequences. The two dimensions of vulnerability (occurrence and severity) are used to characterize the instruments and actions:

- Preventive instruments
  - They aim to reduce the likelihood of occurrence of a disaster.
  - They act on the causes and reduce the probability of occurrence.
- Protective/mitigation instruments
  - They aim to limit the consequences of a disaster.
  - They act on the consequences without taking into account the occurrence.

The goal of all reduction methods is summarized in ▶Fig. 5.2, where we can observe that risks which are considered unacceptable (with the visual help of the risk matrix, see Chapter 4, Section 4.8) have to be brought down to an acceptable or at least tolerable level. Four areas could be focused upon:

- When severity and occurrence are low then we should concentrate on periodically reviewing the situation in order to keep this low level of risk.
- When severity is low and occurrence is high, good housekeeping is necessary.
- When severity is high and occurrence is low, contingency plans should be adopted.
- When both severity and occurrence are high then we should manage actively the risk in order to bring it to a safer level.

We note in the middle zone of the matrix a diagonal intermediate region between the unacceptable zone (needing immediate action) and the acceptable zone. This zone – the "tolerable zone" – is the one where the "as low as reasonably achievable" (ALARA) principle occurs (meaning that either we have to prepare for mitigation measures or we tolerate the risk knowingly). ALARA which means making every reasonable effort to decrease vulnerabilities in order that the residual risk shall be as low as reasonably achievable or practicable (in the latter case, the principle is called "ALARP", see Chapter 3, Section 3.5.2 on Societal risk). This concept was published in the UK in 1974, in the Health and Safety at Work etc. Act (3).

The risk reduction process can be described by ten successive steps:

1. Build a multidisciplinary team that includes specialists in different subjects and sectors, users and a moderator. Sometimes a contribution from someone

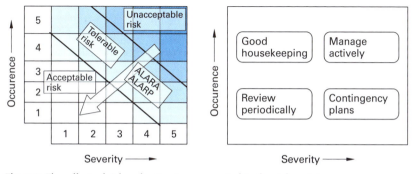

Fig. 5.2: The effect of risk reduction is represented in the risk matrix.

"innocent" (with no or limited experience) is desired. He asks questions no one would have thought of.
2. Identify and define the risks to be reduced (hazard source, situation, requirements, etc.)
3. Analyze the process and operating procedures (description, interaction, objectives, need, constraints, etc.)
4. Generate ideas by brainstorming (a priori, all ideas are good). The multidisciplinary nature of the team reveals all its importance. It is crucial not to discriminate ideas at this stage; it is the moderator's role to guarantee this. From this step will come an accepted solution, discriminating against it too early would be prejudice, especially without selection criteria.
5. Establish criteria to support the evaluation of the suggestions and ideas, a decision help and a referential for the different group members.
6. Evaluate all alternatives [costs (direct and indirect), risk reduction, feasibility, collateral effects, acceptation, opportunity, new hazard apparition, etc.]
7. Retain risk reduction alternatives based on criteria.
8. Implement the decided modification (definition of the person in charge, schedule and deadlines and which necessary means and resources).
9. Information and training – this step is often forgotten, but is crucial. Communication and training is an inevitable need in order for the users to accept and use corrective or improvement measures during a process or activity change. Without communication and training, the will or acceptance of the people who are concerned by the result, cannot be guaranteed.
10. Control, it is indispensable to evaluate the new situation, evaluate the new hazards and risks that have appeared during the implementation of the measures, check the acceptance of the measure by the users and validate the project.

> Risk reduction encloses all strategies whose goal is to limit the risks and damages related to a specific domain. Risk reduction is the promotion of health and assets. It is an integral part of the risk management process of identifying, assessing, and controlling risks arising from operational factors and making decisions that balance risk cost with mission benefits.

5.2 Prevention

Prevention is an attitude and/or a series of measures to be taken to avoid the degradation of a certain situation (social, environmental, economical, technological, etc.) or to prevent accidents, epidemics or illnesses. It has the following characteristics:

- Prevention limits risk with measures aiming to prevent it by eliminating or decreasing the occurrence probability of a hazardous phenomenon. It can be without effect on the severity of potential disasters.
- Preventive measures act upon the causality chain that leads to a loss (a break at a given chain location), e.g., the prohibition of smoking in a room containing flammable materials.
- The effective prevention measures must be based on careful analysis of the occurring causes.

Prevention measures imply:

- Organizing actions for preventing professional risks, information and training.
- Implementing an organization with adapted means.
- Taking into account changes that are likely to happen (new products, new work rhythms, etc.).
- The improvement of existing situations.

The nine principles of prevention include:

1. *Avoid* risks: remove the hazard or the exposure to it.
2. *Assess* risks that cannot be avoided: assess their nature and importance, identify actions to ensure safety and guarantee the health of workers.
3. *Fight* risks at the source: integrate prevention as early as possible, from the design of processes, equipment, procedures and workplaces.
4. *Adapt* work to man: design positions, choose equipment, methods of work and production to reduce the effects of work on health.
5. *Consider the state* of technological developments: implement preventive measures in line with the technical and organizational developments.
6. *Replace* the hazardous by what is less hazardous: avoid the use of harmful processes or products when the same result can be obtained with a method with less hazards.
7. *Plan* prevention integrated in a coherent package: i) technique, ii) work organization, iii) working conditions, iv) social relations, v) environment.
8. Take *collective protection* measures and give them priority over individual protective measures: use of personal protective equipment (PPE) only to supplement collective protection or their defaults.
9. Give *appropriate instructions* to employees: provide them the necessary elements for understanding the risks and thus involve them in the preventive approach.

Preventive measures are developed, taking into account the additional costs generated over the risk. They constitute the *prevention plan*.

5.2.1 Seveso Directive as prevention mean for chemical plants

After the emotion induced by the release of dioxin in Seveso, Italy in 1976, European member states introduced a directive on the control of major-accident hazards involving dangerous substances. On June 24 1982, the Seveso directive asked the member states and enterprises to identify the risks associated to industrial activities and the necessary measures to face them. The Seveso directive was modified several times and its field of action was progressively enlarged, specifically after the Schweizerhalle (Basel, Switzerland) accident in 1986.

The framework of this action is from then on the 96/82/CE Directive, also called the Seveso II Directive, regarding the control of hazards linked to major accidents involving hazardous substances. Amendments (2003/105/CE), which modified the 96/82/CE (Seveso II) directive, were published in the official EU journal on December 31 2003 (4). The new directive strengthened the concept of major-disaster prevention by requiring a company to include the implementation of a management system and organization (or safety system management) proportionate to the risks inherent to the plant. In 2010, the European Commission adopted a proposal for a new directive on the control of major-accident hazards involving dangerous substances that replaced Seveso II, including the global harmonized system for chemicals (GHS). Meanwhile, the Seveso-III European Directive (2012/18/EU) (5) has been published, introducing the GHS for the classification of hazardous substances.

But what is the procedure to change legislation concerning major accidents within Europe? Vierendeels et al. (6) provide the answer, as illustrated in ▶Fig. 5.3.

Some background information is given here to explain ▶Fig. 5.3. Research, debates and analyses on the prevention and protection of human and environment take place on a regular basis, as a "standard procedure" to continuously improve existing regulations. This long-term knowledge (upper-left corner of ▶Fig. 5.3), based on reliable scientific evidence such as expert-knowledge, scientific literature, new

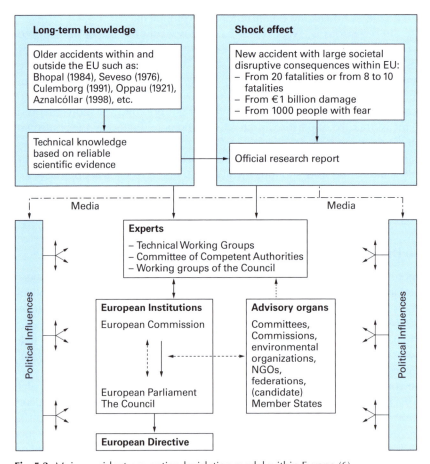

Fig. 5.3: Major accident prevention legislation model within Europe (6).

insights in managing safety and managing chemical products and processes, new best available practices or techniques, international conferences and seminars, etc. lead to changing regulations. This can be considered the "standard procedure" within the EU to change and improve existing legislation. But there are other, "non-standard", factors that lead to new legislation within Europe. The full process is therefore described here.

The European Commission is responsible for writing a proposal and sending it to the European Parliament. Once written by the Commission, a proposal is made public and a debate arises amongst all stakeholders, such as industry federations and non-governmental organizations (NGOs). Advice from different parties (see "Advisory organs" in ▶Fig. 5.3) is provided to the European Commission, officially as well as in more indirect ways (i.e., through influencing and lobbying activities even before the first official text is written). The European Commission subsequently sends an "influenced proposal" to the European Parliament and the Council, where it is discussed by the politicians. A lot of contacts exist between the Parliament and the Council via the so-called Coreper, which is a Committee of Permanent Representatives. The proposal can either be accepted or amended by the Parliament. If amendments are made to the proposal, it is sent back to the European Commission. The Commission then takes all remarks, comments, etc. into account and writes a second proposal. This process of amendment by the Parliament can be repeated once. Then the Coreper is responsible for making a draft of a compromise document. If this fails, the proposal is definitely rejected. If the Coreper succeeds in making a compromise text, the Commission may send this compromise Proposal to the Parliament and the Council for a third reading. A European Directive or Amendment then finally comes into force after an agreement (by votes) is settled between politicians of the European Parliament. If in this final stage there is no majority for the proposal, it is definitely rejected.

As the Seveso legislation concerns and includes highly specific and technical topics, the European politicians are advised by experts in various working groups. In and between these working groups, as well as between the working groups/experts and the European institutions, a continuous reciprocal exchange of information and data takes place. This can be found in the middle of ▶Fig. 5.3. Experts are categorized into three working groups: a) technical working groups (TWGs), b) the Committee of Competent Authorities (CCA), and c) working groups of the Council. In TWGs, technical topics that should be defined more accurately by the legislator or which should be technically elaborated more in-depth in the regulations, are discussed. Findings and recommendations of TWGs are reported to the CCA. The CCA, composed of member states representatives, discusses the Seveso legislation implementation and prepares the texts to be used by the European Commission to write a proposal (i.e., for changing the legislation). The working groups of the Council are composed of government officials of the European Union and of the member states' ministries. These government representatives seek advice from technical experts of their own country.

Although the Seveso legislation is extremely technical (concerning chemical products, their characteristics and tiers, etc.) and the technical experts from the different working groups contribute in an essential way with their expertise and their knowledge, political influences are present throughout the entire legislation process. For example, the experts are influenced by a variety of pressure groups, local politicians and private companies. Local politicians influence European politicians, who will meet with the CCA. During these negotiations, lobbying activities are ever present. Such lobbying can be seen as some kind of cost-benefits analysis, which ultimately leads to a societal optimum between ecology, safety, health and economy. The media and the press also play an important role in this regard and may/will influence politicians in particular (7).

Next to the continuously updating and improving of regulations via the experts/technicians and the working groups, a "shock effect" (via a major accident) may also lead to changing Seveso regulations. The left upper-part of ▶Fig. 5.3 illustrates the long-term continuous improvement of the Seveso legislation, whereas the right-upper-part denotes the shock effect that could possibly lead to changes. But when can an accident be labeled as a shock effect? One important element is the number of fatalities accompanying the accident. Mac Sheoin (8) indicates that the number of people killed has an impact on the speed of the legislation change process. Research revealed that 20 casualties will cause a shock effect for certain. However, societal disruption can already take place from eight to 10 fatalities, in coherence with other factors. A parameter that has a certain influence on the perceived impact of an accident (and thus indirectly on the level of pressure to change the regulations) is the type of victims: employees of the company suffering the accident, rescue workers (such as firemen, emergency rescuers, civil protection, police force), or nearby civilians. The first group of victims (employees) has the least impact, whereas the last group (nearby civilians) has the greatest impact. Although non-European major accidents sometimes cause similar devastating effects to human health and environment, they are not taken into account in the major accident prevention legislation process in a similar ("shock effect") way as those major accidents happening within European borders. Non-European accidents not only require time to investigate (and the incident investigation reports are not immediately recognized by European Authorities), but time is also needed to compare the different societal and organizational cultures and climates before objective and adequate conclusions can be drawn. They are certainly taken into account in the major accident prevention legislation change process in the long run. The older accidents, which did not play a role at first but became influential at later stages in time, are denoted in ▶Fig. 5.3 in the upper-left corner. They belong to the "long-term knowledge" influential factor and are taken into account by various experts in direct and indirect ways. Research reports from various "older accidents" are used in official accident research reports of "new accidents", usually to pinpoint similar causes and the lack of lessons learned from previous accidents (e.g., Aznalcóllar for Baia Mare, Culemborg for Enschede, and Oppau for Toulouse).

The number of fatalities is clearly not the only cause of societal disruption and hence it is not the sole "shock effect" inducement factor for legislative changes. The Baia Mare accident e.g., did not include any fatalities; nonetheless this accident was cited by the 2003 Seveso amendment. Research by Vierendeels et al. (6) indicate that fear for large-scale diseases in the long-term caused by heavy metals and cyanide in the aquatic environment led to societal disruption in this particular case. A visible and traceable disposal of chemical substances in air, water, or soil does make people afraid of major long-term health effects, especially if carcinogenic substances are involved.

The cost of an accident may also induce legislative changes. One billion Euros seems to be a psychological minimum tier that leads to more or different regulations. Accident costs could be property damage, production losses, losses because of image problems, external effects, etc. The latter external effects – the financial consequences of an accident imposed on society – are especially pivotal in this regard.

Media attention because of the shock effect parameters of a major accident will always lead to societal pressure on the political decision-makers. In fact, societal disruption (caused by the shock effect parameters) will influence the legislative process as pictured in ▶Fig. 5.3. The result is ad hoc legislative action, based on the research reports' and research commission's conclusions and recommendations.

5.2.2 Seveso company tiers

Company sites are classified "Seveso" in terms of quantities and types of hazardous materials they host. There are two different thresholds: the "Seveso low" ("tier 1") and the "Seveso high" ("tier 2") threshold.

The overflow threshold is calculated based on the type of products and their risk phrases. The thresholds are present in the directives' annexes. For example, the threshold for combustive substances (Risk phrases R7, R8 and R9) are 50 tons and 200 tons. A company that can store on its site 40 tons of oxygen (combustive R8) and 50 tons of peroxides (combustive R7) is classified "Seveso low tier" or "Seveso tier 1" as the mass of combustive substances is superior to 50 tons but inferior to 200 tons. Without the peroxides, this organization would not be concerned by the directive.

As well as the thresholds of combustive substances (R7, R8 and R9), the directive also suggests different thresholds for explosive substances (R1-R2-R3), flammables (R10), for easily flammable substances (R11), for highly flammable substances (R12), for substances toxic for humans (R26-R27-R28), toxic for the environment (R50-R51/53), etc.

As indicated in the major accident prevention legislation model, the directive took into account the different accidents that happened throughout history in different countries following the awareness of the hazards of major accidents. After this, some accidents became the basis of the addition of amendments to the initial directive. As reminder, some major industrial accidents are listed below:

- The Flixborough Catastrophe (UK, 1974), explosion following the rupture of a cyclohexane conduit, 28 dead [9].
- The Seveso disaster (Italy, 1976), dioxin release following a runaway reaction induced by a malfunctioning heating. No deaths, 200,000 affected by dioxins [10–11].
- The Bhopal catastrophe (India, 1984), one of the largest industrial disasters, releasing methyl isocyanate following an unwanted reaction with water, tens of thousands of deaths (3787 immediate and 15,000 thereafter) [12–14].

- Romeoville (Illinois, USA, 1984), refinery explosion caused by an amine absorption unit, 17 dead (15).
- Mexico City (Mexico, 1984). Boiling liquid expanding vapor explosion or BLEVE of LPG, followed by a domino effect in a petrol terminal, 650 dead and more than 6400 wounded (16).
- Schweizerhalle disaster (Switzerland, 1986), a chemical warehouse catches fire. The Rhine gets polluted as a side effect. The pollution is perceptible in several countries (17).
- Phillips catastrophe (Pasadena, TX, USA, 1989), polyethylene plant explosion, 23 dead, 314 injured (18).
- Enschede disaster (the Netherlands, 2000), fire and explosion of a fireworks warehouse, 22 dead, 950 wounded (19).
- AZF factory explosion in Toulouse (France, 2001), caused by ammonium nitrate (20). The 2003/105/CE directive modified in 2003 the threshold limits of several substances, including that of ammonium nitrate 31 dead, 2442 injured.
- Buncefield disaster (UK, 2005), because of the domino effect it affected several petrol stocking warehouses (21) 43 injured and billions of euros of financial losses.
- Georgia disaster (Port Wentworth, GA, USA, 2008), dust explosion in a sugar refinery, 13 dead, 42 wounded (22).
- Foxconn disaster (Chengdu, China, 2011), dust explosion in iPad2 polishing workshops, four dead, 15 wounded (23).

Some elaborated and explained examples of such accidents are provided in Chapter 11.

> Prevention is an attitude and/or a series of measures to be taken to avoid the degradation of a certain situation (social, environmental, economical, technological, etc.) or to prevent accidents, epidemics or illnesses. It acts mainly on the likelihood of occurrence and the causality chain, trying to lower the probability that an event happens. Prevention actions are also intended to keep a risk problem from getting worse. They ensure that future development does not increase potential losses.

5.3 Protection and mitigation

Protection and mitigation consist of all measures reducing consequences, severity or development of a disaster.

There are two types of protection measures (safeguards):

- *Before the event ("protection")*: reducing the size of the object of risk exposure when an event occurs.
- *After the event ("protection by means of mitigation")*: are usually emergency measures to stop the damage accumulation or counteract the effects of the disaster.

There is also a difference between active and passive protection. Let us take the example of fire protection:

- Build firewalls = passive protection.
- Establishing a system for detecting and/or sprinkler = active protection.
- Establish an evacuation plan + drills = active protection.

Remedial actions reducing the impact of a proven risk are developed, taking into account extra cost versus occurring risk. Together they form the *emergency plan*.

Safety barriers (see ▶Tab. 5.1) are all the adopted available measures regarding conception, construction, and exploitation modalities including internal and external emergency measures, in order to prevent the occurrence and limit the effects of an hazardous phenomenon and the consequences of a potential associate accident (24).

There are two types of safety measures regrouping:

1. *Preventive* measures (▶Tab. 5.2), which are the first barriers to put into place because they aim to prevent a risk by reducing the occurrence likelihood of a hazardous phenomenon.
2. *Protection* measures (▶Tab. 5.3): aim to limit the spread and/or severity of consequences of an accident on vulnerable elements, without modifying the likelihood of occurrence of the corresponding hazardous phenomenon. Protection measures can be implemented "as a precaution", before the accident, e.g., a confinement. We can distinguish within the protection measures, limitation barriers (enfeeblement or mitigation) who aim to limit the effects of a hazardous effect, without modifying its occurrence likelihood (e.g., an explosion vent).

Tab. 5.1: Example of safety barriers.

Safety barrier		Definition	Example
Technical	Passive safety devices	Unitary elements whose objective is to perform a safety function, without an external energy contribution coming from outside the system of which they are part and without the involvement of any mechanical system	• Retention basin • Rupture disk
	Active safety devices	Non-passive unitary elements whose objective is to fulfil a safety function, without contribution from energy from outside the system of which they are part	• Discharge valve • Excess flow lid
	Safety instrumented systems	Combination of sensors, treatment units and terminal elements whose objective is to fulfil a safety function or sub-function	• Pressure measure chain to which a valve or a power contractor is linked
Organizational		Human activities (operations) which do not include technical safety barriers to oppose the progress of an accident	• Emergency plan • Confinement
Manual action systems		Interface between a technical barrier and a human activity to ensure the success of a safety function	• Pressing on an emergency button • Low flow alarm, followed by the manual closing of a safety valve

Tab. 5.2: Examples of preventive measures.

Preventive measures	Example
Fire risk elimination	Using non-flammable materials
Limiting hazardous functioning parameters	Continuous following and automatic control of the functioning parameters classified as critical
Installing isolation, blocking and restriction devices	Rendering flammable liquids inert, locking down electrical equipment
Installing a failsafe after a failure	Use of electrical fuses and circuit breakers
Reducing the likelihood of breakdowns and errors	Oversizing important elements, using redundancy
Substance leak recuperation	Neat floor cleaning.

Tab. 5.3: Example of protection measures.

Protection measures	Human	Material	Organizational
Passive: which act by their presence alone	• Mastering the urbanisation • Seclusion rooms • Escape ladders	• Firewalls • Storage underground • Retentions basin	• Emergency plan existence
Active: which act only with a specific human or material action	• Following orders • Wearing personal protective equipment • Using portable extinguishers	• Sensor activating safety systems: cut off valves, water curtain • Valves, rupture disks	• Operator training • Activating the emergency plan • Implementing crisis cell

Instrumental failsafes are made up of the three following elements:
- Detection elements (sensors, detectors) whose role is to measure the drift of a parameter by signal emission.
- Elements ensuring the gathering and treatment of the signal coming from the implementation of the safety logic (programmable electronic system, relay, etc.) in view of giving orders to the associated actuator.
- Action elements (actuators, motors) whose role is to put the system into a positive safe state and to maintain it that way.

The availability is measured by the probability of failure on demand or PFD (see also Chapter 3, Section 3.4). Examples of available barriers are given respectively in ▶Tab. 5.4 and 5.5 for prevention and protection technical barriers.

> Protection and mitigation consist of all measures that reduce consequences, severity or development of a disaster. Mitigation means taking action to reduce or eliminate long-term risk from hazards and their effects.

Tab. 5.4: Probability of failure on demand (PFD) of some technical prevention barriers.

Prevention barrier	PFD	Commentary
"All or nothing" valve	5×10^{-3} to 10^{-1}	Valid with complete and planned maintenance
Prevention or overpressure/ sub-pressure protection valve	10^{-3} to 10^{-1}	Safety valve function (no-opening when asked)
Rupture disk	10^{-3}	Typical values

Tab. 5.5: Probability of failure on demand (PFD) of some technical protection barriers.

Prevention barrier	PFD	Commentary
Fix fire-fighting equipment	10^{-2}	Typical value if tested regularly
Catalytic gas detection with associated alarm	5×10^{-3}	Valid with tests and calibration once every 2–3 months for a redundant mechanism
Alarm triggering	10^{-3}	Typical value (probability of failure on siren request)
Confinement against explosion/toxic risks	0	Value close to 0

5.4 Risk treatment

The treatment of risks is the central phase of engineering risk management. Thanks to action taken at this stage, the organization can concretely reduce the (negative) risks it faces. These actions should act on the hazard, the vulnerability of the environment, or both, when possible. It requires an organization to identify, select and implement measures to reduce risks to an acceptable level.

Three specific steps are involved in the treatment of risk:

1. Identification of potential measures under the prevention, preparedness, response and recovery domains.
2. Evaluation and selection of measures.
3. Planning and implementation of chosen measures.

Risk treatment is hence described as a selection and implementation process of measures that are destined to reduce (negative) risks.

The ISO 27001 (25) norm imposes the implementation of an analysis method and risk treatment capable of producing reproducible results. Most of the available methods rely on tools that enable an answer to these constraints and aim to treat the steps described in ▶Fig. 5.4.

As observed, the risk treatment answers to five consecutive questions are:

1. What are the issues? Defines the list of sensitive process.
2. Why and what to protect? Gives a list of sensitive assets.
3. From what to protect? Enumerates the list of threats.
4. What are the risks? Proposes a list of impacts and potential.
5. How to protect from them? Resumes the list of safety measures to be implemented.

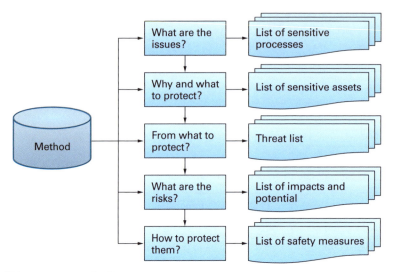

Fig. 5.4: Risk treatment method.

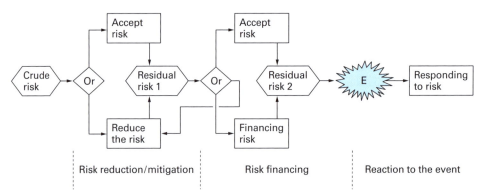

Fig. 5.5: Risk treatment process, (E stands for event).

Putting this methodology in a flowchart, at every step, following questions can be asked, as depicted in ▶Fig. 5.5:

- Is the risk acceptable?
- Should the risk be reduced?

It seems evident that when an event had happened, there is no reason anymore to ask these questions. We must react to the risk, to its realization and consequences.

It is convenient to emphasize that during risk treatment, a more detailed risk analysis can be necessary in order to dispose of the necessary information for an appropriate identification, evaluation, and selection of the measure to be taken. We must count on an adequate analysis level to assure the taken measure really treats the causes of the risks.

On the basis of the accessible information about the risks and the existing control measures, and depending on the priority functions of the established treatments, the first step will be to determine the measure to be implemented to reduce risks.

Thinking should be done, following a logical sequence. As an example:

1. *Evaluate* whether the measure can eliminate or prevent the risk.
2. Then, *analyze* the potential measures to aim to reduce the occurrence likelihood and the intensity of the chance.
3. Finally, *examine* the possible means to lower the environment vulnerability and, at the same time, reduce the consequences.

This identification of potential measures should be carried out while looking at the (/any) measure adapted to a particular risk, and taking into consideration that the measure can also be applied to most other risks. The objective must be to dispose of the best possible combination to optimize the resources and assure an adequate management of all the risks to which the community or the organization is exposed to. It must also target the implementation of situations where several risks are interacting in one and the same environment.

Finally, this treatment option evaluation should also consider the different legal, social, political and economic factors. It should also be part of a perspective that contemplates the establishment of measure in the short- and long-term.

▶Tab. 5.6, 5.7 and 5.8 expose selected categories of safety measures or barriers susceptible to be evaluated and implemented in the risk treatment step.

Tab. 5.6: Examples of safety measures for different categories.

Measure categories	Examples
The legal and normative requirements	Adoption of laws, regulations, policies, orders, codes, standards, proceeded to certification, etc. intended to govern or to supervise the management of a chance, a risk or a domain of wider activity
The consideration of the risks in the town and country planning and development	Rules governing the land-use in the exposed zones, the standards of immunization, maximal densities of land-use, rules of compatibility, prescriptions towards materials and specific techniques of construction to increase the resistance of infrastructures and buildings, the realization of technical studies inside an exposed zone, development evacuation paths, etc.
The elimination or the reduction of the risk at the source	Modification of industrial processes, use of a less hazardous product, modification of routes transportation, installation of equipment or realization of works allowing to decrease risks, to limit its probability of occurrence or to decrease its potential intensity, establishment and application of directives and procedures of reduction of the risk, etc.
The rehousing of the people and the displacement of the exposed goods	Population, residences, infrastructures, etc. (it is a means of last recourse that intervenes generally when a risk is considered unacceptable by a community and when the other measures of prevention and preparation do not represent a valid option)

Tab. 5.7: Examples of safety measures for structural actions.

Measure categories	Examples
Infrastructures, developments and equipment intended to avoid the appearance	Protection wall of a reservoir of hazardous materials, works to prevent the release of an avalanche, rocks to avoid a landslide, etc.
Mechanical or physical means to reduce the probability of occurrence or the intensity	Retention basin for rainwater, intensification of tanks to avoid leaks of hazardous materials, fast severing mechanisms for leaks and emergency stop in the industrial installations, etc.
Inspection and maintenance programs	Measures to prevent the development of conditions convenient to the appearance of risks and to assure the preservation of safe conditions in the execution of risky activities: order in the equipment, state of buildings and infrastructures, hygiene and public health, etc.
The financial and fiscal capacities	Dissuasive measures to prevent or to limit the risk by the imposition of a penalty (tax or higher price rate), incentive measures to encourage the realization of actions allowing risks reduction, the subscription to an insurance to obtain a compensation in the eventuality of losses consecutive to a disaster, etc.
Research and development programs and activities	Research and development on varied subjects such as fast alert systems, surveillance and forecast mechanisms, methods and tools of risk appreciation, materials and techniques of construction allowing an increase in the resistance of buildings and infrastructures, etc.
Public awareness programmes, risk communication and the population preparation	General campaigns to raise awareness, communication on the nature and the characteristics of risks, the exposed territory, the predictable consequences, the measures taken to avoid the disaster, the means with which the citizens have to protect themselves, records to follow in case of disaster, etc.

Tab. 5.8: Examples of modalities to assure intervention and restoring.

Measure categories	Examples
Modes and procedures of alert and mobilization	Measures intended to warn the population, the responsible authorities and the participant in an emergency situation or a disaster and to put into contribution the resources necessary for the management of the situation, etc.
Help measures for the population and goods protection and the natural environment	Search and rescue, evacuation, put under cover, health care, control, protection of the goods and the natural environment, etc.
Measures aiming to preserve the essential services and operations (continuity of the operations) and the protection of economic activities	Drinking water, energy, transport, telecommunications, financial services, emergency services, health system, food supply and essential governmental services, support for companies, resumed by the activities and return on conditions of public health: levying of protective measures, cleaning and reassurance of places, restoring of the services, etc.

(Continued)

(*Continued*)

Measure categories	Examples
Help measures for the population	Services to the disaster victims, the psycho-social care, the management of the needs of the whole community, support following a disaster: financial, psychological, technical support, etc.
Modes and mechanisms of public information Procedures for the experience feedback	Communications at the time of and following a disaster: instructions to the citizens, the relations with the media, etc. Production and broadcasting of reports on the causes and the circumstances of the event, the holding of sessions of evaluation of the operations following a disaster or following an exercise, an analysis of the answer to the disaster and the measures of reduction of the risks of setting up, etc.
Exercise programs	Exercises of alert, traffic of information, mobilization, activation of a center of coordination, coordinated management, operational and general evaluation with or without deployment, etc.
The administrative and logistic modalities	Agreements, procedures and administrative directives for the mobilization of the human, material and informative resources, the acquisition of the material resources, the preparation of the installations: centers of coordination, accommodation, etc.
The follow-up and revision capacities of the level of preparation	Check of the functioning and the maintenance of the installations, the equipment and the intervention equipment, programs of information of the participants, the procedures of update and periodic revision of the measures, the periodic check of the level of preparation: reports, questionnaire of self-assessment, audit, etc.

Risk reduction is primarily a matter of removing or containing the source of the risk. There are a number of alternatives for preventing exposure, including capping and paving, and these could be applied where appropriate. An analysis of the actual risks posed in particular situations would be required on a site-specific basis to determine the appropriate risk reduction method. Once the treatment measures are defined they still need to be challenged and assessed for different criteria as depicted in ▶Tab. 5.9.

Risk treatment consists not only of trying to reduce or mitigate the risk, but also on how to finance the risk or potential losses and finally how to react when an event occurs. There is a need to integrate the principles and practices of sustainability with the principles and practices of risk treatment. Only by adopting a sustainable approach can effective, equitable and long-term approaches to treating risks and building resilience be developed.

5.5 Risk control

If during a risk assessment it is observed that the work system is not safe enough, or that the risk is too high for the group of people examined, then appropriate measures must be sought to eliminate or reduce risks. In assessing once again the risk with the selected measure, we check whether the chosen measure effectively reduces the risk. It must also be checked at that time whether the implementation of new measures of protection involves additional or new hazards.

Tab. 5.9: Some criteria for assessing risk treatment options [adapted from (26)].

Criteria	Question to be answered
Cost	Is this option affordable? Is it the most cost-effective?
Equity	Do those responsible for creating the risk pay for its reduction?
Timing	Where there is no man-made cause, is the cost fairly distributed?
Leverage	Will the application of this option lead to further risk-reducing action by others?
Administrative efficiency	Can this option be easily administered, or will its application be neglected because of difficulty of administration or lack of expertise?
Continuity of effect	Will the effects of the application of this option be continuous or merely short-term?
Compatibility	How compatible is this option with others that may be adopted?
Jurisdictional authority	Does this level of government have the legislated authority to apply this option? If not, can higher levels be encouraged to do so?
Effects on the economy	What will be the economic impacts of this option?
Effects on the environment	What will be the environmental impacts of this option?
Risk creation	Will this option itself introduce new risks?
Risk reduction potential	What proportion of the losses due to this risk will this option prevent?
Political acceptability	Is this option likely to be endorsed by the relevant authorities and governments?
Public and pressure group reaction	Are there likely to be adverse reactions to implementation of this option?
Individual freedom	Does this option deny basic rights?

If that is the case, these hazardous phenomena have to be added to the list of already noticed hazardous phenomena and a new risk appreciation must be performed. ▶Fig. 5.6 presents an iterative process of risk reduction.

Risk control measures are based on:

- *Prevention* measures whose role is to reduce the probability of the occurrence of feared events that are the source of hazard for damaging targets.
- *Protection* measures whose role is to protect targets against the effects of such things as heat flow, pressure or projectiles, which are associated with the release of dangerous phenomena.
- *Mitigation* measures whose role is to limit the effects of the appearance of feared events.

Prior to analysis, the limits of the system that forms the business or process to consider have to be defined. We should also define precisely what is in the system, and what is therefore taken into account in the identification of hazards, and what lies outside that system and the various modes of operation.

Large areas or processes should be divided into smaller components. If a sector of activity or a process includes a whole line of production composed of several facilities, then the different sectors or partial processes should, when possible, correspond to a process phase. The interfaces between the whole system and the environment, as well as the interface between some parts of the sector or parts of the process, have to be well-defined and highlighted.

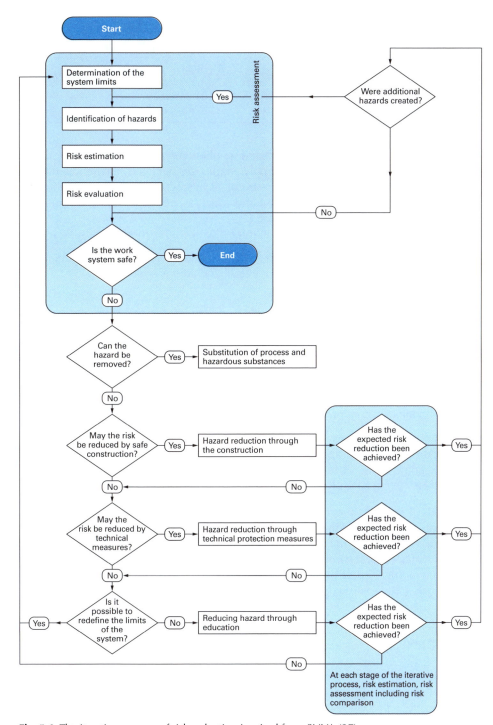

Fig. 5.6: The iterative process of risk reduction inspired from SUVA (27).

Tab. 5.10: Exposure and priority of preventive actions.

Exposure process	Priority of preventive actions
1. Emission	1. At the source of hazard
2. Transmission in the environment of the target	2. At the interface between the target and the source.
3. Exposition	3. At the target

It is necessary to specify the type of hazardous phenomena under consideration, and to indicate to whom and to what they apply (employees, facilities, environment, etc.). It is also important to clarify if there are eventual interactions with the neighbouring facilities that must be taken into account and which aspect does not need inspection (construction static, process chemistry, etc.).

We must make a particular effort not to forget the operating modes of a system as:

- Normal operation: The facility fulfils the function for which it was designed.
- Special operations: Prepare, convert, install, and adjust, errors, clean.
- Maintenance:
  - Control (measure, control, and record); determine the real state and compare with the predicted state.
  - Maintenance (cleaning and maintenance); measures to conserve the desired state.
  - Replacement (replacement and improvement) restore to the previous state.

Let's take chemistry as example. The choice of preventive action follows a parallel hierarchy to the exposure process (see ▶Tab. 5.10):

1. Priority should always focus on the search for measures that limit the use or emission of harmful agents.
2. Next come measures that prevent the spread of these agents to workers.
3. And last, personal protective measures and community protective measures.

> Risk control includes methods by which firms evaluate potential losses and take action to reduce or eliminate such threats. It is a technique that utilizes findings from risk assessments (identifying potential risk factors in a firm's operations, such as technical and non-technical aspects of the business, financial policies, and other policies that may impact the well-being of the firm), and implementing changes to reduce risk in these areas.

5.6 STOP principle

When looking for appropriate solutions to safety problems, we have to, first of all, clarify whether the hazardous phenomenon can be deleted by replacing certain substances and some dangerous processes. If it is not possible to delete the hazard by improving the construction work or by using less hazardous substances, we must then proceed with technical and organizational measures and as a last resort, measures relative to people.

5.6 STOP principle

The STOP (strategic, technical, organizational and personal measures[1]) principle underlines this approach by giving priority to:

- The measures in the following order:

  (i) *S* measures: *strategic*, substitution of processes or substances giving a less hazardous result (e.g., substituting, eliminating, lowering, modifying, abandoning, etc.); abandon process or product, modify final product.

  (ii) *T* measures: *technical* protection against hazardous phenomena that cannot be eliminated, lowering the likelihood of occurrence of an event and reducing the spread of the damage, (e.g., replacing, confining, isolating/separating, automating, firewall, EX zones, bodyguards, etc.).

  (iii) *O* measures: *organizational* modifications of the work, training, work instructions, information concerning residual risk and how to deal with it (e.g., training, communicating, planning, supervising, warnings signs, etc.)

  (iv) *P* measures: *personal*, relative to people (e.g., personal protective equipment, masks, gloves, training, communication, coordinating, planning, etc.)

- The hierarchy of the priorities in the following order:

  (i) Acting at the *source*: deleting the risk (substituting product or process, in situ neutralization), limiting leak risks (re-enforcing the system, lowering the energy levels), predictive measures (rupture disk, valves) and surveillance (integrity and functionality of the system, energy levels).

  (ii) Acting at the *interface* (on the trajectory between the source and the target): limiting the propagation (active barriers/passive barriers), catching/neutralizing (local or general ventilation, air purification, substance neutralization), people control (raising barriers, access restrictions, evacuation signs) and surveillance [energy levels in the zone, excursions or deviations (alarms)].

  (iii) Acting at the *target*: lowering the vulnerability (personal protective equipment selection, special training), reducing exposure (e.g., automation), reducing the time (job rotation) and supervising (individual exposure, biological monitoring, medical survey, correct PPE use and following rules).

In general, we must combine measures to obtain the required safety. It is important that the choice of safety measures enables the reduction of the likelihood and severity of the hazardous events. To make this choice, we must not only take into account the short-term costs, but also the long-term profitability calculations. Once the priorities have been established, it is possible to determine the correct method to master each of the identified risks. These methods are often regrouped in the following categories:

- Elimination (including substitution).
- Engineering measures.
- Administrative measures.
- Individual protective equipment.

---

[1] We also refer to the P2T model from chapter 3. The analogy between "TOP" and "P2T" is obvious: P2T stands for Procedures (or organizational), People (or Personal), and Technology (or Technical).

Tab. 5.11: The STOP table.

	At the source	At the interface	At the target
Measures **S** (strategy)	• Substitution • Change process	• Automation, telemanipulation • Room subdivision	• Criteria for selection of licensed operators
Measures **T** (technical)	• Reactant production or use in continuous mode • Safety relief valves	• Fumes extraction process enslaved • Physical access restrictions	• Selection and purchase of PPE and CPE
Measures **O** (organizational)	• Response instructions	• Extraction of fumes manually controlled • Access restrictions markup	• Prescription for PPE and CPE • Organization of first aid
Measures **P** (personal)	• Education/training of the process operation	• Information/instruction on the process hazards	• Instruction for the use of PPE

PPE, personal protective equipment, CPE, community protective equipment.

▶Tab. 5.11 presents a recap of the ordering of measures and the considered environment, illustrated by few examples for each category. Directions of approach are from top to down and then from left to right.

Eliminating the hazard is the most favorable approach when reducing risks; substitution is interesting as long as it does not generate new hazards. No hazard, no risk. In the STOP principle, the elimination and substitution phases are included in the strategic measure S. They are, however, rarely possible in practice, thus eliminating and substituting may sometimes not be applicable.

Let's take the example of substituting a solvent (benzene) in a chemical reaction by a solvent with little toxicity (1, 2-dichlorobenzene). Thus the toxic effect of benzene is eliminated and replaced by a lower toxicological effect. From this point of view, the problem is over. However, during the synthesis with the new solvent, side-products have appeared whose toxicity could cause other dramatic effects. So what must we do? One problem has been replaced by another. If another hazard of equal importance had not been introduced, the suggested solution would have won the vote. This example shows that the strategic measure is not always applicable in the field. An easy measure, from the risk management point of view, would be to say that the synthesized product is no longer interesting, so no more synthesis, no more solvent and the problem is solved. But can we live without this product? This is the strategic question!

The principle of the STOP concept is a continuous and consecutive approach and is summarized in the following ▶Fig. 5.7. Note that the S part (strategical) is missing as it consists of substituting, eliminating or changing a process. Strategic measures are not dependent on the process. Often this principle is shortened to TOP because strategic or substitution measures are already been taken whenever possible.

The definition of "hazard" (which is to be eliminated or substituted) can be established with the help of two questions, that is, "Why?" and "How?", in relation to the product and the product necessity.

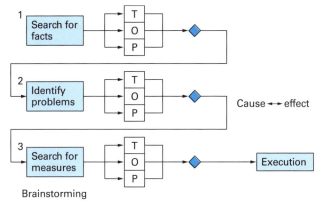

Fig. 5.7: The TOP iterative process.

The principle of the method in three steps for each TOP measure is depicted in ▶Fig. 5.7:

1. Researching facts.
2. Deducing the relative problems.
3. Searching for the appropriate measures.

As for every principle, there are not only pros or only cons. However we will not discuss the method itself, although quite logical, but rather the different types of measures it recommends. The advantages and inconveniencies of the different types of measures (STOP) are summarized in ▶Tab. 5.12.

Note that in practice, personal protection measures are put into place before the technical and organizational measures. This happens for many different reasons, including, costs, delays, implementation simplicity, loss of responsibility, to have no time or take no time to analyze the situation, the simplicity, etc.

Many organizations have invested heavily in personnel, processes and technology to better manage their risk. But these investments often do not address the strategy and

Tab. 5.12: STOP principle pros and cons.

	Pros	Cons
Measures **S** (strategical)	• Cancel or reduce the considered hazard • Intervene at the beginning of the process	• In case of a substitution, it is possible to create other hazards or risks • Deletion needing a strategic decision
Measures **T** (technical)	• Fixed • Difficult to bypass	• Costs • Deadlines
Measures **O** (organizational)	• Quick • Moderate costs	• Controllability • Easy to bypass
Measures **P** (personal)	• Quick • Moderate costs • Simple implementation	• Controllability • Acceptability • Convenience • Omission

processes that should be implemented. To successfully turn risk into results, we need to become more effective at managing scarce resources, making better decisions and reducing the organization's exposure to negative events by implementing the four-level steps comprising strategic, technical, organizational and personal aspects.

5.7 Conclusion

Risk treatment/reduction is the selection and implementation of appropriate options for dealing with risk. It includes risk avoidance, reduction, transference and/or acceptance. Risk reduction is used as a preferred term to risk termination. Often, there will be residual risk, which cannot be removed totally as it is not cost-effective to do so. Risk acceptance is sometimes referred to as risk tolerance.

Risk treatment involves identifying the range of options for treating risks, assessing these options and preparing and implementing treatment plans. The risk management treatment measures may be summarized as:

- Avoid the risk: decide not to proceed with the activity likely to generate risk.
- Reduce the likelihood of harmful consequences occurring: by modifying the source of risk.
- Reduce the consequences occurring: by modifying susceptibility and/or increasing resilience.
- Transfer the risk: cause another party to share or bear the risk.
- Retain the risk: accept the risk and plan to manage its consequence.

Risk reduction should be integrated into a cost-benefit analysis. It can play a pivotal role in advocacy and decision-making on risk reduction by demonstrating the financial and economic value of incorporating risk reduction initiatives into aid planning.

Risk reduction is appropriate in risks that have high cost, human or environmental implications, and where their probability to occur is relatively higher so that the business organization may not be able to solely bear the burden of the uncertainty and is forced to strategically minimize on the probability of occurrence and severity of impact of the risk.

References

1. Operations Directorate Public Safety Canada (2011) Public safety of Canada, National Emergency Response System. Ottawa, Canada: Operations Directorate Public Safety Canada.
2. Frost, C., Allen, D., Porter, J., Bloodworth, P. (2001) Risk management overview. In: Operational Risk and Resilience. Oxford, UK: Butterworth-Heinemann.
3. Health and Safety at Work etc. Act (1974) (c. 37). legislation.gov.uk, latest revision 2012.
4. Council of the European Communities (2003). Directive 2003/105/EC. Official Journal of the European Union, L 345, 31/12/2003, 97–105.
5. Directive 2012/18/EU of the European Parliament and of the Council of 4 July 2012 on the control of major accident hazards involving dangerous substances, amending and subsequently repealing Council Directive 96/82/EC.

6. Vierendeels, G., Reniers, G., Ale, B. (2011) Modeling the major accident legislation change process within Europe. Safety Sci. 49:513–521.
7. Zoeteman, B.J.C., Kersten, W., Vos, W.F., Voort, L. van der, Ale, B.J.M. (2010) Communication management during risk events and crises in a globalised world: predictability of domestic media attention for calamities. J. Risk Res. 13:279–302.
8. Mac Sheoin, T. (2009) Waiting for another Bhopal. Global Social Policy 9:408–433.
9. Venart, J.E.S. (2007) Flixborough: A final footnote. J. Loss Prevent. Proc. 20:621–643.
10. Cardillo, P., Girelli, A., Ferraiolo, G. (1984) The Seveso case and the safety problem in the production of 2,4,5-trichlorophenol. J. Haz. Mat. 9:221–234.
11. Bertazzi, P.A. (1991) Long-term effects of chemical disasters. Lessons and results from Seveso. Sci. Total. Environ. 106:5–20.
12. Jasanoff, S. (1988) The Bhopal disaster and the right to know. Social Sci. Med. 27:1113–1123.
13. Chouhan, T.R. (2005) The unfolding of the Bhopal disaster. J. Loss Prevent. Proc. 18: 205–208.
14. Eckerman, I. (2011) Bhopal Gas catastrophe 1984: causes and consequences. In: Encyclopedia of Environmental Health, pp. 302–316. Amsterdam, The Netherlands: Elsevier.
15. Dennis, P., Nolan, P.E. (1996) Historical survey of fire and explosions in the hydrocarbon industries. In: Handbook of Fire & Explosion Protection Engineering Principles for Oil, Gas, Chemical, and Related Facilities. Burlington, MA: William Andrew Publishing.
16. Pietersen, C.M. (1988) Analysis of the LPG-disaster in Mexico City. J. Haz. Mat. 20:85–107.
17. Giger, W. (2009) The Rhine red, the fish dead—the 1986 Schweizerhalle disaster, a retrospect and long-term impact assessment. Environ. Sci. Pollut. Res. 16:98–111.
18. U.S. Chemical Safety and Hazard Investigation Board (2007) Refinery Explosion and Fire, Investigation Report No. 2005-04-I-Tx, pp. 1–341.
19. World Health Organization (2009) Public Health Management of Chemical Incidents. Geneva, Switzerland: WHO.
20. Dechy, N., Bourdeaux, T., Ayrault, N., Kordek, M.A., Le Coze, J.C. (2004) First lessons of the Toulouse ammonium nitrate disaster, 21st September 2001, AZF plant, France. J. Haz. Mat. 111:131–138.
21. Buncefield Major Incident Investigation Board (2008) The Buncefield Incident 11 December 2005, The final report. J. Loss Prevent. Proc. 1: 2a and 2b.
22. U.S. Chemical Safety and Hazard Investigation Board (2009) Sugar Dust Explosion and Fire. Investigation Report No. 2008-05-I-GA, pp. 1–90.
23. Apple (2012) Apple Supplier Responsibility 2012 Progress Report, pp. 1–27. Cupertino, CA: Apple Inc.
24. Sklet, S. (2006) Safety barriers: Definition, classification, and performance. J. Loss Prevent. Proc. 19:494–506.
25. International Organization for Standardization (2005) Norm n° ISO/IEC 27001.
26. Foster, H.D. (1980) Disaster Planning: The Preservation of Life and Property. New York: Springer-Verlag.
27. SUVA (2001) Méthode Suva d'appréciation des risques à des postes de travail et lors de processus de travail, Document n°66099, pp. 1–47.

6 Event analysis

In this chapter we will use the term "event analysis" to mean accident, incident or near-miss analyses. Human and organizational factors are an important cause of accidents. As the design of electro-mechanical equipment becomes more and more safe, the causes of accidents are more likely to be attributed to human and organizational factors. Event analysis is conducted in order to discover the reasons why an accident or incident occurred, and to prevent future incidents and accidents. Investigators have long known that the human and organizational aspects of systems are key contributors to accidents, indicating the need for a rigorous approach for analyzing their impacts. Safety experts strive for blame-free reports that will foster reflection and learning from the accident, but struggle with methods that require direct technical causality, also they do not consider systemic factors, and they seem to leave individuals looking culpable.

It is hard to keep your cool when trying to analyze the objective causes of an incident or accident and to produce a common report. This is where the difficulty arises, as it is hard to find a common language that is limited to the objectivity of the facts. Indeed, when an accident happens, a sensitivity aggravating climate is installed and thinking gives place to arguing and everyone tries to find who or what is responsible without even trying to understand anything. The event is often considered as the result of a combination of unfortunate circumstances: it is such and such at fault ... it could have been worse, how unlucky ... inhibiting any ulterior analysis. This phenomenon is illustrated in ▶Fig. 6.1 – there is a huge divergence between facts, their interpretation and the suppositions.

An event analysis method is therefore needed to guide the work, aid in the analysis of the role of human and organizations in accidents, and promote blame-free accounting of accidents that will support learning from the events and thus becoming a proactive process.

We may still raise the question: "Why analyze accidents?" The answers are multiple and could be listed as:

- To understand by analyzing the objective causes related to the accident.
- To prevent a recurrence.
- To determine which measures will improve safety.
- To act by implementing adequate solutions.
- To show employees that safety and health protection must be taken seriously.
- To communicate and therefore cool down debates.
- Savings for companies.
- To meet regulations.

The common denominators among the major analytical techniques are that:

- We do not seek for responsibilities but for solutions.
- An accident is always a combination of several factors.
- We should retain only facts, no judgments nor interpretations.

Fig. 6.1: Rumors and facts.

- The analysis is interesting only if it leads to the implementation of solutions.
- The analysis of accidents is a collective work.

> We should never forget that near-misses, even if they have not concretized in an incident or accident, are crucial to investigate, in order to prevent the realization of the risk leading to damages.

6.1 Traditional analytical techniques

Traditional analytical techniques deal mainly with the identification of accident sequence, and seek unsafe acts or conditions leading to the accident. Such techniques include the sequence of events, multilinear events sequencing, root cause analysis, among many other available techniques. The methods presented below represent only a portion of the available accident analysis techniques. The primary objectives are to:

- Learn to identify and stop destructive behaviors.
- Understand the benefits of the observation process.
- Understand how to positively communicate observations.
- Build on employee empowerment.
- Learn the monitoring and tracking system.
- Develop a prioritized observation and strategy plan.

6.1.1 Sequence of events

This technique was originated by Herbert William Heinrich in 1929 (1), based on the premise that accidents result from a chain of sequential events, metaphorically like a line of dominoes falling over. When one of the dominoes falls, it triggers the next one, and the next ... but removing a key factor (such as an unsafe condition or an unsafe act) prevents the start or the propagation of the chain reaction. (See also Chapter 3, Section 3.3.)

In the newest version of the domino theory model, five labeled dominoes – i) lack of control by management, ii) basic causes, iii) symptoms, iv) incident, v) loss of property and people – form the basis of the chain of dominoes. Each domino represents one event. A row of dominoes representing a sequence of events leading to the mishaps is lined up. When one domino falls (when an event in the sequence occurs), the other dominoes will follow (1,2). However, should a domino in the sequence be removed, no injury or loss will be incurred. Note that such an analysis is usually confined to accidents happening in the exact sequencing.

James Reason (3) (see also Chapter 3) proposed a similar model to that of the domino theory model and the accident analysis framework through the incorporation of the pathogen view on the causation of accidents (the likelihood of an accident is a function of the number of pathogens within the system). Together with triggering factors, the pathogens will result in an accident when the defenses in the system are breached. The model is based on a productive system, which comprises of five elements: i) the high-level decision-makers, ii) line management, iii) preconditions, iv) productive activities and v) defenses. The pathogens in a typical productive system originate from either human nature or an organization's strategic apex (high-level decision-makers). The associated types of pathology to each respective productive element are fallible high-level decisions, line management deficiencies, the psychological precursors of unsafe acts, and inadequate defenses (3).

6.1.2 Multilinear events sequencing

Multilinear events sequencing was conceived by Benner (4). Basically, this technique charts the accident process. The model is based on that the first event to create unbalance in the system constitutes the start of a chain of events that ends in the damage or injury. The accident sequence is thus described as an interaction between various actors in the system. A description of the accident sequence constitutes the starting point for identification of the situation that can explain why the accident occurred. Every event is a single action by a single actor. The actor is something that brings about events, while actions are acts performed by the actor. A time-line is displayed at the bottom of the chart to show the timing sequence of the events, while conditions that influence the events are inserted in the time flow in logical order to show the flow relationship as depicted in ▶Fig. 6.2 (1). With this chart, countermeasures can be formulated by examination of each individual event to see where changes can be introduced to alter the process.

6.1.3 Root cause analysis

Root cause analysis (RCA) is a process designed for use in investigating and categorizing the root causes of events with safety, health, environmental, quality, reliability and

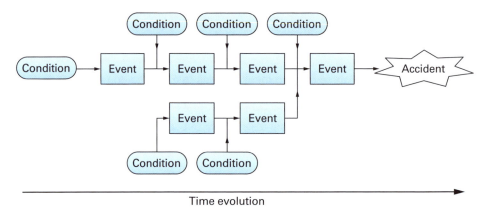

Fig. 6.2: General events sequencing chart.

production impacts. The term "event" is used to generically identify occurrences that produce or have the potential to produce these types of consequences. Simply stated, RCA is a tool designed to help identify not only what and how an event occurred, but also why it happened. Only when investigators are able to determine why an event or failure occurred can they specify workable corrective measures that prevent similar future events. Understanding why an event occurred is the key to developing effective recommendations.

The RCA process involves four steps (5):

1. Data collection and gathering
   - Without complete information and an understanding of the event, the causal factors and root causes associated with the event cannot be identified.
2. Causal factor charting
   - The causal factor chart is simply a sequence diagram with logic tests that describes the events leading up to an occurrence, plus the conditions surrounding these events.
3. Root cause identification
   - After all the causal factors have been identified, we can begin root cause identification. This step involves the use of a decision diagram called the "root cause map" to identify the underlying reason or reasons for each causal factor. The map structures the reasoning process to help with answering questions about why particular causal factors exist or occurred.
4. Recommendation generation and implementation
   - Following identification of the root causes for a particular causal factor, achievable recommendations for preventing its recurrence are then generated and must be implemented.

There are many analytical methods and tools available for determining root causes to unwanted occurrences and problems (5). Useful tools for RCA are, e.g., the "five whys" (6), the Ishikawa diagrams, also called Fishbone diagrams (7), or the FMEA.

> All incident/accident or near-misses analyses tend to discover the main causes that led to the event. Causes may be unsafe acts, unsafe conditions or technical failures. They are generally not self-standing but they link together either in sequential or bijective mode. The sequential interconnection is important in order to correctly analyze the event.

6.2 Causal tree analysis

The causal tree accident analysis method elaborated by the INRS (Institut national de recherche et de sécurité, France) (8), is based on an original work initiated by the European Coal and Steel Community and was experimented for the first time in a practical way in 1970 in the iron mines of Lorraine. This method aims at being situated beyond the debates and the opinions. It offers a way to analyze fine circumstances that have led to an incident/accident, and to transform the causes of these incidents/accidents into predictable facts, hence leading to prevention. It is an investigation and analysis technique used to record and display, in a logical, tree-structured hierarchy, all the actions and conditions that were necessary and sufficient for a given consequence to have occurred.

Causal tree analysis provides a means of analyzing the critical human errors and technical failures that have contributed to an incident or accident in order to determine the root causes. It is a graphical technique that is simple to perform and very flexible, allowing for mapping out exactly what we think happened rather than being constrained to an accident causation model. The diagrams developed provide useful summaries for inclusion in incident and accident reports that give a good overview of the key issues.

It has to be noted that this method has many similarities with the fault tree analysis (FTA). However, FTA is focusing on failures determination (technical, process) but the causal tree is mainly looking for accident causes.

6.2.1 Method description

Tree structures are often used to display information in an organized and hierarchical fashion. Their ability to incorporate large amounts of data, while clearly displaying parent-child or other dependency relationships, also makes the tree a very good vehicle for incident investigation and analysis. The combination of the tree structure with cause-effect linking rules and appropriate stopping criteria, yields the *causal tree*, which is one of the more popular investigation and analysis tools in use today.

Typically, it is used to investigate a single adverse event or consequence, which is usually shown as the top or right item in the tree. Factors that were immediate causes of this effect are then displayed below it or on the left, linked to the effect using branches. Note that the set of immediate causes must meet certain criteria for necessity, sufficiency, and existence. Proof of existence requires evidence.

Often, an item in the tree will require explanation, but the immediate causes are not yet known. The causal factor tree process will only expose this knowledge gap; it does not provide any means to resolve it. This is when other methods such as change analysis or barrier analysis can be used to provide answers for the unknowns. Once the unknowns become known, they can then be added to the tree as immediate causes for the item in question.

The method has four main steps:

1. Search for facts:
   - Without any judgment.
   - Without interpreting.
   - By treating each fact one by one.
   - One fact must be measurable and/or photographable.
2. Build the tree:
   - Begin with the last fact.
   - Develop branches by asking the three following questions:
     - What is the direct cause that provoked this?
     - Was this cause really necessary for the occurrence of that fact?
     - Was this cause sufficient to provoke the event?
3. Search for measures:
   - Begin with the first fact.
   - Search measures for each fact.
   - Accept all ideas.
4. Define measures:
   - Efficiency, use.
   - Measures not displacing a risk.
   - Simple measures, sustainable.
   - Measures complying with laws and regulations.

The overall process could be schematized as depicted in ▶Fig. 6.3, which shows that building a comprehensive tree is not sufficient if it does not lead to solutions and corrective measures implementation. In order to avoid the repetition of a similar accident it is also crucial to install a follow-up procedure.

6.2.2 Collecting facts

A fact is an action or event that really happened. The main aspects are:

- **Form**: collection of hot circumstances and additional elements.
- **Retain a fact list**: directly linked to the accident, validated by the team members.
- **Objectives**: provide an opportunity to identify and establish the circumstances that led to an event.
- **Prerequisites**: a report should be made straight after the accident, all team members should have been on the event location.
- **Qualifications**: be curious, careful; choose words with precision and accuracy.

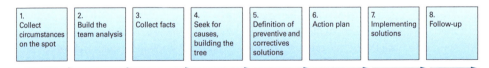

Fig. 6.3: Overall process of the causal tree.

They are several rules to follow that are vital when collecting facts in order that no judgments or interpretations are brought to the analysis.

- No judgments.
- Do not make any assessment.
- Identify only the facts.
- Note one independent fact at the time.
- Facts are measurable and/or photographable, in simple terms.
- Describe the facts.
- Facts must be unambiguous.
- Do not write down feelings, except from an injured person when they are expressed spontaneously.
- Do not use adjectives (beautiful, dangerous, many, little, etc.)

Let us take an illustrative example to discover that facts are not always easy to retain. Read the description of the situation and then try to answer to the following questions.

The car body repairer John Bely arrives at the garage from the 5th avenue. In the court, a worker shouted in Spanish with big gestures, looking up. Then, he enters the room and starts to change tires on a car. Near the porch, a young bearded man speaks tenderly to someone who has long hair and wears jeans.

Proposals	Answers		
	True	False	?
John Bely is garage mechanic.			
There are only four persons in the court.			
John Bely is on the 5th avenue.			
The contractor has heard a worker shouting.			
One of John Bely's workers was changing tires.			
A Spanish worker was in the court.			
The worker shouting in the court was speaking with a colleague located on an upper floor.			
He shouted to alert his colleagues of the arrival of John.			
A bearded man and a woman were talking.			

Sometimes, it will be necessary to interview witnesses to gather facts. A witness can be defined by any person who has information relating to the incident or accident. One should be aware that evidence obtained through witness testimonies and statements should be considered to be very fragile. It is very rare for any witness to be able to recall clearly all the details of an event (before, during, and after). Realistically, each witness will likely have a slightly different description of what he or she observed. The differences between witness statements can be explained by several reasons, such as:

- No two individuals see details or objects and remember them the same way.
- Points-of-observation vary from witness to witness.
- Witnesses are different in both technical and personal backgrounds.
- Witnesses are interested in self-preservation and the protection of friends.

- Witnesses differ in their abilities to rationalize what they have seen and articulate an explanation or description.

Some basic tips for the investigator to start an interview (2), are:

- Explain who he/she is.
- Explain why the accident is being investigated.
- Discuss the purpose of the investigation (e.g., to identify problems and not to determine fault or blame).
- Provide assurances that the witness is not in any danger of being compromised for testifying about the accident.
- Inform the witness of who will receive a report of the investigation results.
- Ensure witnesses that they will be given the opportunity to review their statements before they are finalized.
- Absolutely guarantee the privacy of the witness during the interview.
- Remain objective – do not judge, argue, or refute the testimony of the witness.

To gather as much relevant information as possible, the questions should focus on *who, what, when, where,* and *how* aspects of the incident or accident. Such focus will preclude the possibility that the witness will provide opinion rather than fact. While the *why* questions can be asked at the end of the interview, it is a straightforward rule that the majority of the interview should be based on facts and observations and not on opinions.

At least the following areas should be covered during an interview of a witness:

1. *What* was the exact or approximate time of the event?
2. *What* was the condition of the working environment at the time of the event; before the event; after the event (e.g., temperature, noise, distractions, weather conditions, etc.)?
3. *Where* were people, equipment, and materials located when the event occurred; and *what* was their position before and after the event?
4. *Who* are the other witnesses of the event (if applicable) and *what* is their job function?
5. *What* and/or *who* was moved from the scene, repositioned, or changed after the event occurred?
6. *When* did the witness first become aware of the event and *how* did he or she become aware?
7. *How* did the response or emergency personnel perform, including supervisors and outside emergency teams?
8. *How* could the event have been avoided?

The first five questions establish the facts of the event. This information is required in order to establish an understanding of the facts as they happened or as they were perceived, but also to validate the credibility of the witness testimony. Question 6 is an indication of the point in time during the event when the witness became a witness. It is obvious that such a question may lead to vital information about the cause of the accident; e.g., if the witness would claim that he or she became aware that something was wrong because of a strange sound or visual happening prior to the actual event. The purpose of question 7 is to obtain feedback on the effectiveness of the organization's existing loss control procedures. The last question serves to fix a point of reference at the end of the investigation. An opinion on the prevention of the event may actually provide a direction from which the investigator can proceed when providing recommended solutions, prevention and protection measures, mitigation measures, corrective actions, etc.

6.2.3 Building the tree

Once the immediate causes for the top item in the tree are known, then the immediate causes for each of these factors (see ▶Fig. 6.4) can be added, and so on. Every cause added to the tree must meet the same requirements for necessity, sufficiency, and existence. Eventually, the structure begins to resemble a tree's root system. Chains of cause and effect flow upwards or from left to right of the tree, ultimately reaching the top level. In this way, a complete description can be built of the factors that led to the adverse consequence. Remember that the three question to be answered when identifying facts are (▶Fig. 6.4):

1. What is the direct cause that provoked this?
2. Was this cause really necessary for the occurrence of that fact?
3. Was this cause sufficient to provoke the event?

Once the facts are determined, there are three possibilities to connect them together as described in the following recap (▶Fig. 6.5).

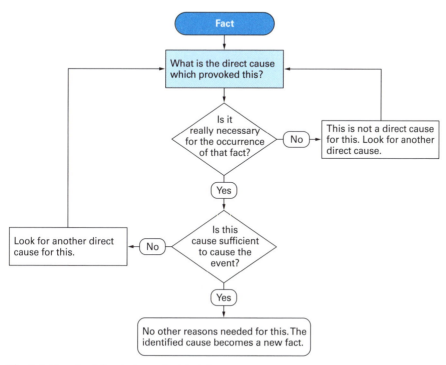

Fig. 6.4: How to define independent facts?

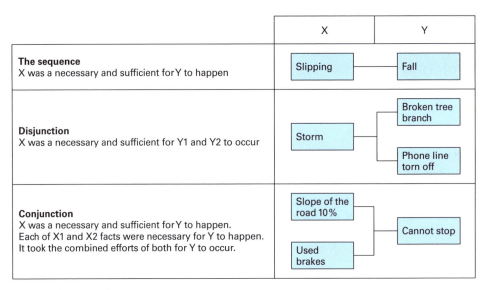

Fig. 6.5: The types of connectors.

The remaining task is to build the complete tree, either from the top down or from right to left depending on the tree size. Before illustrating the method with an example, let us try to find the right connection of facts through a small illustration.

In this brief case, four facts have been identified, one of them being the top event or the final fact.

1. The faucet is left open.
2. The tub overflows.
3. The siphon is blocked.
4. The bathroom carpet is full of water.

The first step is to determine the final event: in this case, fact number 4, "the bathroom carpet is full of water" is the top event.

They are several possibilities to interconnect the remaining facts with the top event as illustrated in ▶Fig. 6.6, however only one is the reflection of what happened. Four facts, three possible combinations – which one is the correct solution, A, B or C?

Let us find the correct solution. We should begin by asking what was the direct cause that led to "The bathroom carpet is full of water", according to the procedure described in ▶Fig. 6.4. The answer is that only fact number 2 "The tub overflows" is sufficient for fact 4 to happen. So fact 2 is sequentially linked to fact 4. Thus solution C cannot be the correct one.

Next we start with fact number 2 and ask the same question as before: what was the direct cause that led to the tub overflowing? The answer is that both fact number 1 "The faucet is left open" and fact number 3 "The siphon is blocked" are necessary for fact 2 to happen. So facts 1 and 3 are linked by conjunction to fact 2. This excludes solution A, revealing that the correct causal tree is solution B.

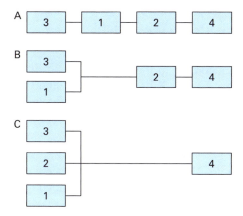

Fig. 6.6: Different possible causal trees.

6.2.4 Example

Now, we are ready to illustrate the causal tree analysis by evaluating the following accident description:

An employee was ordered to sort the bottles by glass color. Customers can return the empty bottles in crates. The delivery truck will take empty bottles from the customer and the driver visually checks if the crates are filled. The customer does not receive any reimbursement for the missing bottles. The employee takes out every bottle from the crate and puts it on the conveyor belt installation. The bottles that are not corresponding to the specified glass are set aside in a special crate for this purpose. By pulling out a defective bottle with a broken neck, the employee injured his right hand causing a severe tendon cut of the right thumb. The collaborator will be out of work for 2 months duration.

When working on the facility, wearing gloves is mandatory as specified in the policy of the company. Consequently, employees receive personal protective gloves.

Firstly, we have to collect the facts:

1. Workers must take the bottles out of crate to be sorted by type.
2. Mr X does not wear protective gloves.
3. The customer is not refunded for broken and/or missing bottles.
4. Protective gloves are not designed for the specific task – hazard.
5. The customer puts the number per supplier in crates without separating by content.
6. Mr X holds the broken bottle.
7. There is a broken bottle in the crate.
8. Mr X has a reflex movement.
9. The driver does not control the bottles when taking them from the customer.
10. Mr X does not meet the guidelines of wearing gloves.
11. Mr X does not see that the bottle is broken.
12. Mr X has a significant injury to his right hand.

13. The customer puts broken bottles in the crate.
14. There is an absence of regulation, "Control of crates by the driver".
15. There is a lack of control by the supervisor.

Secondly, we must build the tree according to the aforementioned rules. It seems obvious that the final event is number 12 "Mr X has a significant injury to his right hand". Building the tree leads to the final causal tree presented in ▶Fig. 6.7.

6.2.5 Building an action plan

Defining measures is easier when the tree is built. Safety features (safety barriers, corrective measures for both prevention and protection) must be provided as barriers in response to the initiating event. They generally aim to prevent, as far as possible, the initiating event as the cause of a major accident. It is usually sufficient to stop one branch of the tree in order to avoid the occurrence of the final event. They are summarized in the *action plan* whose objective it is to determine the measures by answering the following questions:

- What are the chosen measures to stop the repetition of an event?
- What is the measure implementation deadline?
- Who is responsible for their implementation and meeting the deadline?

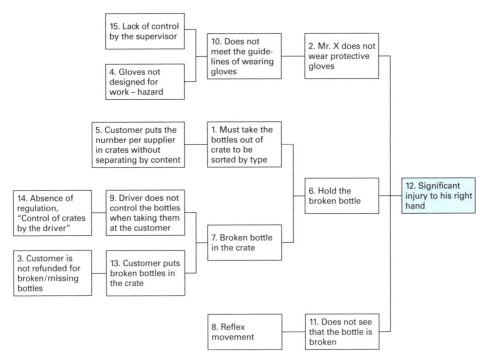

Fig. 6.7: Causal tree from the mentioned example.

The number of solution propositions has to be large enough to build a coherent action plan. It is essential to acquire the skills required to implement measures. The measures are chosen according to the following criteria:

- Efficiency, use.
- No risk displacement.
- Extent consistent with law.
- Effective solution.
- Simple solution (strong acceptance by individuals).
- Eliminating the root causes of the event.
- Possible use of the solution in other areas.

When choosing measures, one must ensure that all the influences included in the causal tree are counteracted by the selected measures. The gathered proposed solutions should be classified depending on their efficiency level. Combinations of measures should be preferred and eligible. For example: technical modification of an apparatus + organization adjustment + personnel and superior formation. *The farther the neutralized influences are from the fact, the more effective they are.*

In the example from ▶Fig. 6.4 we can imagine to choose between facts 4 and 15 and 10 in order to ensure that adequate protective gloves are worn. It is also possible to imagine that the driver looks carefully to identify broken bottles – facts 9 and 13 and fact 7. They are no limits in defining measures; they should only be adequate, economically sustainable, effective and implemented.

6.2.6 Implementing solutions and follow-up

Once the action plan has been established, solutions must be implemented and a follow-up and control must be performed. Communication with personnel is essential as it enables us:

- To value the study made by the team.
- To inform employees about what actually happened.
- To inform employees about the anomalies.
- To inform employees about the retained and adopted solutions.

> The causal tree method allows us, using a graphical representation, to investigate the root causes and subsequent facts that led to an event. It deserves not only the incident/accident analysis but also a teaching purpose to avoid the repetition of such an event by taking appropriate measures. Corrective measures and actions must be based on a careful analysis in order to implement adequate remedies.

6.3 Conclusions

Why should an accident analysis be conducted? The ultimate reason for conducting an accident analysis is to avoid future injuries through the identification of facts. The accident analysis process should not be used to place blame. No person wants to get hurt. We need to conduct a thorough accident and incident analysis to identify the multiple causes of the accident. Then, based on the facts, develop solutions. This activity is designed to stop damaging infrastructure and stop human suffering.

A thorough accident and incident analysis will determine multiple causation and potential trends by department or area of responsibility. With the facts and data gathered in the process of the analysis, corrective measures can be developed and similar injuries will be reduced or eliminated.

A side effect is that accident analysis could be largely used as an efficient educational tool for the training and empowered awareness to safety. It is the main tool in enterprise safety partnerships for the development and the political prolongation of planned preventions, conceived as an element of enterprise management.

References

1. Ferry, T.S. (1977) Modern Accident Investigation and Analysis, 2nd edn. Hoboken, NJ: John Wiley & Sons Inc.
2. Vincoli, J.W. (1994) Basic Guide to Accident Investigation and Loss Control. New York: John Wiley and Sons Inc.
3. Reason, J. (1997) Basic Guide to Accident Investigation and Loss Control. Aldershot, UK: Ashgate Publishing Company.
4. Benner, L. (1975) Accident investigations. Multilinear events sequencing methods. J. Saf. Res. 7:67–73.
5. Vanden Heuvel, L.N. (2007) Root Cause Analysis Handbook: A Guide to Effective Incident Investigation, 3rd edn. ABS Consulting, Published by Rothstein Associates Inc.
6. Ohno, T. (1977) Toyota Production System: Beyond Large-Scale Production. Productivity Press. New York: Taylor and Francis Group.
7. Ishikawa, K. (1990) Introduction to Quality Control. Productivity Press. New York: Taylor and Francis Group.
8. Monteau, M. (1974) Méthode pratique de recherche des facteurs d'accidents. Principes et application expérimentale, Rapport n° 140/RE INRS.

7 Crisis management

We often hear different statements after a major event happened such as:

- "Crises only happen to others"
- "If that should happen, we will improvise"
- "A crisis is unpredictable, so why should we get organized?"
- "People always exaggerate the severity and the consequences of a crisis"
- "We are in a crisis every day and we always get out of it"

So would it not make sense to identify the first signs of the crisis, so that when it happens, we know what to do, or when it is over that we know how to recover? Those are the several points that this chapter will try to highlight.

Disasters have served to guide and shape history, and many great civilizations during mankind were destroyed by the effects of disasters.

Swartz et al. (1) identified at least three stages in managers' mindsets regarding crisis management. The authors suggested a potential evolutionary path for crisis management since the 1970s from the initial focus upon technological activities and concerns predominantly with hardware towards a growing interest in the value of business continuity management. They also stated that there is no suggestion that these mindsets are restricted to specific periods, but rather they suggest that each represented the dominant paradigm for a particular decade:

technology mindset (1970s) → auditing mindset (1980s) → value mindset (1990s) → normalization of business continuity management (2000s) → ???

Crisis management as a corporate activity has the fundamental strategic objectives of ensuring corporate survivability and economic viability when business profits and/or continuity are threatened by external or internal potentially destructive events.

Three characteristics can be expressed as separating crises from other unpleasant occurrences:

- *Surprise* – even naturally occurring events, such as floods or earthquakes, do not escalate to the level of crisis unless they come at a time or a level of intensity beyond everyone's expectations.
- *Threat* – all crises create threatening circumstances that reach beyond the typical problems organizations face.
- *Short response time* – the threatening nature of crises means that they must be addressed quickly. This urgency is compounded by the fact that crises come as a *surprise* and introduce *extreme threat* into the situation.

> In this chapter we will consider a crisis as a specific, unexpected, and non-routine event or series of events that create high levels of uncertainty and that are threatening.

7 Crisis management

7.1 Introduction

Damages following an accident can lead to a degraded situation by weakening the target affected by the accident. This target is then exposed to the emergence of other accidents against which it is usually protected. The degradation situation then becomes a crisis situation. A breach in the defence system and it is the whole system which is at threat. This phenomenon is highly dependent on time as illustrated in ▶Fig. 7.1. The crisis is multifaceted and multi-temporal; it can be:

- The result of a succession of past events leading to a creeping degradation.
- The result of a major event, a sort of sudden cataclysm.
- The outcome of future events that will be linked together in cascade.

As observed, at the beginning, threats (represented by arrows) are not reaching the target (the vulnerable resource, process, human, etc.). This represents the normal status, the target is not harmed and safety barriers are fulfilling their job. When the situation deteriorates with time, the target could be reached by the threats, and then damages and/or harms appear. The immediate consequence is a weakening of the target as the safety barriers are either deteriorated or not completely fulfilling their duty. At the end, a chain of accidents allows multiple threats on a weakened target and safety barriers, indicating that the crisis had worsened and that the target will suffer from multiple and more important damages and/or harms (2).

This implies that, following an accident, one must not only repair the noticed damage, but at the same time be more careful and try to prevent other risks whose likelihood of occurrence momentarily increases drastically.

A *crisis* is often defined "as an acute situation, hard to manage, with important and lasting consequences (sometimes harmful)". It can result from an accident or from the normal evolution of a situation. This last point can lead us to think that it would be more correct to consider a crisis as a passing process from one situation to another, e.g., from a critical situation to a catastrophic situation. A crisis assumes a decision-making process and immediate action to get out of it. *A crisis is an unusual situation characterized by its instability, forcing a specific governance to be applied in order to get back to a normal status.*

We must not confuse a crisis with an exceptional event or an emergency situation. A crisis is a period of intense instability, more or less long, not anticipated, not always

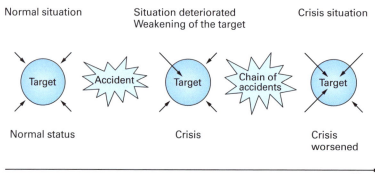

Fig. 7.1: Physical modeling of the time evolution effect on the crisis.

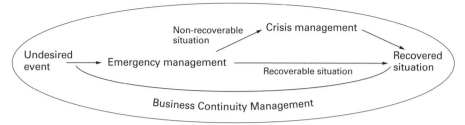

Fig. 7.2: Emergency, crisis, and business continuity management.

visible or known, which causes negative consequences on the environment and its actors. ▶Fig. 7.2 provides an overview of the differences between emergency, crisis and business continuity management.

The components of a crisis management system are composed of:

- Pre-event risk and vulnerability assessment.
- Mitigation and loss control.
- Planning.
- Response management.
- Recovery functions.

The fundamental strategic objective of any organization is to ensure its long-term survivability and economic success. Crisis management is a strategic function that links functions such as risk management, safety management, environmental management, security, contingency planning, business recovery, and emergency response (3).

When an event such as an industrial accident, a toxic release, or a major oil spill occurs, management often finds itself simultaneously involved in emergency management (the specialized response of emergency forces), including: disaster management (the management of the incident and the management and support of the response organization); crisis management (the management of the crises situations that occur as a result of the accident); and business recovery and continuity (the entire management process to deal with the recovery and continuation of the profitable delivery of products and services). In addition, crisis management often includes a strong focus on public relations to recover any damage to public image and assure stakeholders that recovery is underway.

A crisis is often defined as an acute situation, hard to manage, with important and lasting consequences (sometimes harmful). It can result from a triggering accident or from the normal evolution of a situation. It has the following rules:

- Being unprepared is no excuse.
- You know the threats – get ready for them.
- Know in advance before you are asked.
- Admit that you are wing-it-challenged.
- Adopt short key message communication type.
- Beware of the court of public opinion.
- The first 72 hours of any crisis are the crunch time.
- Do not forget, that in a crisis situation time flies.
- Get every help or support you may need.
- Every crisis is an opportunity.

7.2 The steps of crisis management

In reality, managing is predicting. Anticipating and preparing a protocol and crisis management organization is a very big responsibility for the company's management. Managing a crisis and containing the crisis so it does not extend or deteriorate, is a high priority objective of any organization.

In a crisis situation, it is difficult to predict human behavior with regards to the undergone stress and emotional load that can overwhelm an individual. *Anticipation*, the master word of crisis management, is:

- Imagining and studying the different incurred scenarios in the company, by also integrating the known catastrophic scenarios.
- To equip ourselves with "thinking" and "acting" in a structured way when the moment comes, known as the "crisis cell".
- Prepare a minimum of the functioning logistics and procedures.
- Prepare the crisis communication so as not to add an aggravating factor in case of mediatisation.
- Identify in advance the right people: those who have the knowledge and expertise, those who are united and can work in team, those who will not "flinch" when there are decisions to make and finally those whose legitimacy would not be questioned.

Anticipating and preparing for a crisis is answering the following questions:

- How should we react and be organized to face an unpredictable disaster?
- Which crisis management device should be deployed?
- Which collaborators should be implicated? What should they be attributed and what would their responsibilities be?
- Which logistic processes/activities should be planned?
- Which crisis communication should be put into place? And for which addressees?
- Which procedures and which operation should be implemented to reduce crisis impact and proceed with the restarting of the activities?

There are the risks that are known, have been integrated and that are now assumed … and then there are all the other risks, the new ones, the unknown unknowns, the unexpected, which play tricks on our emotions and psychosis, anger … irrationality. This constitutes one of the main difficulties of crisis communication. To make sure the words are not transformed into ills, every used term must be chosen with care and precision by the public powers and political, economical and media actors.

And it is precisely this mixture of uncertainty, anxiety and acceleration that multiplies 10-fold the evolution risk of a "sensitive" situation (or emergency situation) towards a "crisis" situation. This crisis situation is always characterized by its instability, which forces a specific governance to be adopted in order to go back to a normal way of life. By crisis management, we mean a governance mode. In a crisis situation, we are heavily engaged in the future, and in a way that is necessarily more closed than usual. As an extreme example, it is not when an airline pilot approaches critical take-off speed and he notices a motor failure that he will start consultations and negotiations to decide which manoeuvre to make or to perform an emergency brake. What is done at that moment does not translate the instantaneous good will of the individual: it is the anterior decisions that play a large role.

7.2.1 What to do when a disruption occurs

No matter what the quality of the risk management system is, an unwanted and unexpected event will happen sooner or later, as "zero risk" does not exist. As long as the risk is there, identified and quantified, no matter how well-managed, we know the accident may happen. But the moment when it will happen remains, of course, unknown. The same goes for a crisis: it surprises and, under Murphy's Law, some say it always happens at the wrong moment (4).

Mythology is ripe with tragically sad stories. The story of Cassandra is such a mythological tale. Cassandra was princess of the legendary city of Troy; she was the daughter of King Priam and Queen Hecuba. She was a precocious and charming child that easily caught the attention of mortals and Gods alike.

As was very common in Greek mythology, the God Apollo lusted after the beautiful young mortal woman and intended to make her his own. To convince her to give into his advances, he promised to bestow upon her the gift of prophecy and being able to foretell the future.

While Cassandra was obviously flattered that an Olympian God sought her favors, she was not at all convinced that she wanted to take him as lover. Still, unable to resist the gift he offered, she eventually relented.

Apollo took Cassandra under his wing and taught her how to use her prophecies. Once her mentorship was finished, however, Cassandra refused to give her body to Apollo as promised.

Furious at being rejected by a mere mortal, Apollo decided to punish her. However, he could not take back the gift he had already given her. So he leveled a terrible curse upon her head. While Cassandra would still be able to foresee the future, the curse ensured that no one would believe her. Worse than that, they would believe that she was purposely telling lies. True to his word, Cassandra was able to foresee the future for herself and those around her. But every attempt she made to warn people in advance of impending doom was ignored or, worse yet, labeled as an outright lie.

The careful reader immediately sees the analogy with crisis and disaster managers: before anything happens, they are often regarded by others as fantastical, and doom tellers who do not have to be believed. "Nothing ever happened, so why would something happen now?" is often the reasoning. Hence, in the proactive phase, crisis and disaster managers are often treated exactly the same way as Cassandra.

The major steps of crisis management could be expressed as three phases (see ►Fig. 7.3):

1. Preparation: *before*, is composed of:
   - Major risk identification (linked to strategic impact risks, necessitates risk quantification, focuses on critical consequences).
   - Prospective crisis organization – crisis scenarios and continuity plans, crisis levels, ethics and image. In case of a crisis, we can be confronted with ethical problems. It is possible, e.g., that some resources have to be sacrificed in order to save others.
2. Treatment and repair: *during*, is composed of:
   - Triggering of the crisis. It is a matter of wisely triggering the crisis management process. If it is done too early, the system is destabilised, credibility is lost and

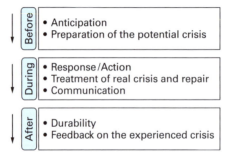

Fig. 7.3: Major steps of crisis management.

we risk being no longer capable of mobilization if the situation really requires it. Too late and the consequences will be dramatic.
- Crisis cell. The establishment of a crisis cell cannot be improvised at the last moment, but should be prepared, the members of the cell having previously participated in one or more simulations. The crisis cell has the mission of evaluating the disaster, taking immediate protection measures for people and the affected facilities. It must also inform the board of directors and the local authorities with assured communication and determine a strategy to get back to a normal situation. The crisis cell is an emergency organization of the service leader to identify a problem, put a contingency plan into action and return as quickly as possible to a normal situation.
- Crisis communication. This should be prepared: the possible attitudes, the choice of speech and tone of the message to send out are predetermined in the crisis scenario. It takes care of the relationship with the media. Note that the media are: quick, simplifying, aggravating and amplifying.
- Repair. Once the crisis reaches its peak and is in its descending phase, repairs can begin. The period preceding the crisis has been able to put into place temporary solutions. Repairs do not stop at these solutions and have the mission of returning back to normal.
- End of the crisis situation. The crisis situation has an end, it is thus necessary to record it. It enables an end to the emergency and surveillance devices that have been put into place in parallel with the crisis. It is a subtle balance between taking the devices away too fast and leaving them in place too long.
3. Memorisation: *after*, is composed of:
   - Learning from the situation, documenting and analyzing the different steps and sequences.
   - Feedback, learn from the situation and improve a new occurrence.
   - Continuity plan updates in order to learn from experience.

When disruptions occur, they are handled in three steps:

1. Response (incident response involves the deployment of teams, plans, measures and arrangements).
2. Continuation of critical services (ensure that all time-sensitive critical services or products are continuously delivered or not disrupted for longer than is permissible).

Fig. 7.4: Generic crisis typology matrix.

3. Recovery and restoration (the goal of recovery and restoration operations is to recover the facility or operation and maintain critical service or product delivery).

Crises and causes of crises can be e.g.: adverse weather, bribery, blackmail, computer breakdown and/or failure, fire, flood, pandemic, IT-problems (computer virus), liability issues, loss of resources, loss of staff, major accident, energy supply problems, reputation problem, sabotage, supplier problem, internet failure, etc. Mitroff et al. (5) developed a matrix for classifying types of resources according to where the crisis is generated and which systems are the primary causes. This matrix provides a starting point for a creative brainstorming session that provides a means of identifying the range of disruptions an organization might experience; see ▶Fig. 7.4.

Hence, different types of crises are, e.g.:

- Internal crises:
  - Product problems, supply chain rupture, process problems, absence of key people, fire, hostility, financial problems …
- External crises
  - Product/services, consumers, suppliers, community, competition, regulation, fire, hostility, stock-exchange movement, natural disasters, …

Rather than deal with individual incidents, "families" of crises may be clustered together and provide a focus for preparations.

> Managing a crisis involves three steps: before, during and after the crisis. The latter being often the most lasting and more time and resources demanding.

7.2.2 Business continuity plan

When the crisis situation occurs it is too late to ask the question of "What to do?", too late to search for information – we must simply act. The principle consists of elaborating different scenarios, looking for failures that could follow an initiating event. *The purpose of developing a business continuity plan (BCP) is to ensure the continuation of the business during and following any critical incident that results in disruption to the normal operational capability.* It must enable us to work in a degraded mode or in a major crisis situation. It is a strategic document, formalized and regularly updated, of planned reactions for catastrophes or severe disasters. Its main goal is to minimize the impact of a crisis or a natural, technological or social catastrophe has on the activity (thus the sustainability) of an enterprise, a government, an institution or a group (6).

The scenarios that are deemed possible will constitute the contingency plan, also called the continuity plan. It focuses on how to keep in service an activity whose usual means of operation have been partially destroyed. It must also plan for measures to limit the damage and prevent it from spreading to other activities.

Disaster recovery and business continuity planning are processes that help organizations prepare for disruptive events. Disaster recovery is the process by which you resume business after a disruptive event. The event might be something huge, like an earthquake, or something small, like malfunctioning software caused by a computer worm.

Given the human tendency to look on the bright side, many business executives are prone to ignoring "disaster recovery" because disaster seems an unlikely event. BCP suggests a more comprehensive approach to making sure you can still do the business, not only after a natural calamity but also in the event of smaller disruptions including illness or departure of key staff, supply chain partner problems or other challenges that businesses face from time to time.

The following steps could be identified:

1. Size up the risks:
   - Assess how potential risks to the business/process will impact the ability to deliver products and services.
   - Differentiate between critical (urgent) and non-critical (non-urgent) organization functions/activities.
   - Develop a list of possible scenarios that could seriously impede the operations.
2. Identify vulnerable target and critical services, assess the risks:
   - Determine who would be affected most by each possible crisis and which areas of operations would be hit the hardest.
   - Critical services or products are those that must be delivered to ensure survival, avoid causing injury, and meet legal or other obligations of an organization.
   - Compile a list of key contacts and their contact information when determining which external stakeholders would be most affected.
3. Formulate the plans – to respond to an emergency:
   - Create plans to address each of the vulnerable targets identified in the previous step. Plans, measures, and arrangements for business continuity.
   - Develop a generic checklist of actions that may be appropriate when an emergency occurs.
   - Test the plan, when crisis occurs it is too late to ask oneself if the BCP will work.
4. Readiness procedures:
   - Determine critical core businesses or facilities.
   - Determine the core processes required to ensure the continuation of the business.
   - Prepare the response (team response, emergency facilities, how to react)

5. Return to business
    - Create plans to return to "business as usual" the ultimate goal in any disaster is to bring operations back into line by rebuilding damaged facilities and reorganizing the work force to cover any losses.
6. Quality assurance techniques (exercises, maintenance and auditing).
    - Review of the BCP should assess the plan's accuracy, relevance and effectiveness. It should also uncover which aspects of a BCP need improvement. Continuous appraisal of the BCP is essential to maintaining its effectiveness. The appraisal can be performed by an internal review, or by an external audit.

BCP is a proactive planning process that ensures critical services or products are delivered during a disruption.

> Critical services or products are those that must be delivered to ensure survival, avoid causing injury, and meet legal or other obligations of an organization. Business continuity planning is a proactive planning process that ensures critical services or products are delivered during a disruption.

7.3 Crisis evolution

In a similar way to a fire, it is in the first few minutes that an exceptional event can become an emergency situation, which can then degenerate into a crisis.

When are we in a crisis?

- On principle, too late!
- Not always when we expect it.
- Never when we would like.
- When the director of the company knows or assumes that the risky event could be serious for the company.
- Following a non-mastered emergency situation.

A crisis, like every other dynamic process, does not evolve in a monotonous way in time (7). We can schematize the different steps in time, as depicted in ▶Fig. 7.5:

We could simplify this model by breaking a crisis up into three stages; pre-crisis stage, acute-crisis stage, and post-crisis stage identified by the circled numbers 1, 2 and 3.

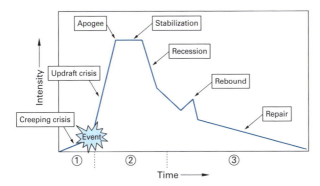

Fig. 7.5: Chronology of a crisis situation.

7.3.1 The pre-crisis stage or creeping crisis

When someone in an organization discovers a critical situation, they usually bring it to the attention of their supervisors/managers. This is known as either the pre-crisis warning or precursor. At this point in time, the critical situation is known only inside the organization and is not yet visible to the general public. When managers are told of the critical situation, their job is to analyze it to determine if it has the potential to become serious. If managers are then comfortable with it and feel it will be resolved without any action on their part, they will not take any action. If, however, they see the critical situation as a serious problem requiring intervention, they will take action to mitigate it. They should then manage it and prevent it from moving into the acute-crisis stage. This is considered a time of opportunity, to turn this from a negative situation into a positive one. The first issue is then to recognize the situation for what it is and what it might become.

7.3.2 The acute-crisis stage

A crisis moves from the pre-crisis to the acute stage when it becomes visible outside the organization. At this point in time, we have no choice but to address it. It is too late to take preventative actions, as any action taken now is more associated with "damage control". Once the problem moves to the "acute" stage, the crisis management team should be activated. The main actions to be taken are:

- Take charge of the situation quickly.
- Gather all the information you can about the crisis and attempt to establish the facts.
- Communicate the story to the appropriate groups that have vested interest in the organization, namely, the media, the general public, the customers, the shareholders, the vendors, and the employees.
- Take the necessary remedial actions to fix the problem.

7.3.3 The post-crisis stage

A crisis moves from the acute-crisis stage to the post-crisis stage after it has been contained. This is when the organization will try to recoup its losses. The organization must show the customer, all shareholders, public authorities and the community that it cares about the problems the crisis has caused. During this stage, the organization must not forget to:

- Recoup any losses – recovery phase.
- Evaluate its performance during the crisis – challenging phase.
- Make any changes that were identified during the crisis – learning phase.

> Crisis evolves with time. It is not a straight or smooth process; it could split into three major time evolution stages: pre-crisis stage, acute-crisis stage, and post-crisis stage.

7.3.4 Illustrative example of a crisis evolution

A crisis is not static, it is dynamic. It evolves following a cycle that is almost always the same, represented in ▶Fig. 7.5.

Let us illustrate this chronology by taking the example of Hurricane Katrina in August 2005 (see ▶Fig. 7.6), with a partial extract from U.S. Department of Health & Human Services and the National Institute of Environmental Health (8,9).

Hurricane Katrina was most destructive hurricane ever to strike the USA. On August 28 2005, Hurricane Katrina was in the Gulf of Mexico where it powered up to a category 5 storm on the Saffir-Simpson hurricane scale. This catastrophe illustrates perfectly the chronology of a crisis as depicted in ▶Fig. 7.5. We can divide the crisis evolution as:

1. Creeping crisis
   - Thursday, August 25, 2005: tropical storm Katrina threatens to gain power the warm waters of the Atlantic and rises to the rank of a category 3 hurricane before striking the southeast coast of Florida.
2. Updraft crisis
   - Friday, August 26, 2005, 04h45: Hurricane Katrina reached the southeast coast of Florida where she has already killed at least two people and deprived hundreds of thousands houses without power.
   - 21h50: Hurricane Katrina killed five people in south-eastern Florida and knocked out power for more than two million people, before finding the strength to touch the warm waters of the Gulf of Mexico.
3. Apogee
   - Sunday, August 28, 2005, 09h25: Residents of New Orleans begin to evacuate the city threatened by Hurricane Katrina which is approaching with renewed vigour.
   - 09h50: The merchants of the French Quarter (Vieux Carré) form New Orleans piled sandbags in front of the front of their stores to protect them from floods that may be caused by Hurricane Katrina.

Fig. 7.6: Satellite picture of Katrina Hurricane [With reprint permission of NOAA (10)].

- 22h15: The winds of Hurricane Katrina has significantly increased steadily, reaching category 5, and blowing at 284 km/h, threatening to cause catastrophic damage.
4. Recession
    - Monday, August 29, 2005, 09h26: Katrina is downgraded to category 4 but it is not excluded that it could strengthen up the land to return to category 5.
    - 12h54: In the eye the east of the hurricane, the Mississippi overflows and floods Slidell and Mobile up to 9 to 10 meters.
    - 13h00: The hurricane is located 112 kilometres from the city; it progresses at a speed of 25 kilometres per hour.
5. Rebound
    - Monday, August 29, 2005, 13h21: The conditions start to improve in New Orleans, but degrade rapidly in Picayune and Springhill.
    - 13h38: There is water seepage in the Superdome, where 10,000 people were present because they could not be evacuated by their own means.
    - 14h04: Ochsner hospital is flooded up to the first floor; patients are discharged into the upper floors.
    - 15h05: the drinking water system is polluted and unfit for consumption.
    - 16h53: Hurricane Katrina was downgraded to a Category 3 hurricane.
6. Repairs
    - Tuesday, August 30, 2005, the extent of major damage in several states made up the provisional invoice over 25 billion dollars, Hurricane Katrina became the most expensive in history and one of the most deadly.
    - Wednesday, August 30, 2005, 13:00: Given the number of damaged or flooded dams, and now with over 90% of homes flooded, the mayor of New Orleans ordered the total evacuation of the last residents, refugees, and ordered the total city closure for at least 6 to 12 weeks; it is likely that the pumping of flooded areas will take months.
    - Emergency services decided not to deal at the moment with the floating bodies and concentrate their efforts on the survivors.
    - Thursday, September 1, 2005, the evacuation of the refugees from the Superdome in New Orleans has been suspended after shots had been fired at military helicopters coming to search for them. The National Guard decided to send hundreds of military police to regain control.

Currently, in 2013, repairs of the damage caused by Katrina are not yet fully completed. The efforts to return to normality continue.

7.4 Proactive or reactive crisis management

As anyone who has been involved in a crisis knows, bad news travels alarmingly fast (especially in our era, where social media is ever more important in society and is used by more and more people). Proactive crisis management involves forecasting potential crises and planning how to deal with them, e.g., how to recover if the computer system completely fails.

Reactive crisis management involves identifying the real nature of a current crisis, intervening to minimize damage and recovering from the crisis.

The more an organization denies its vulnerability, the more its activities will be directed to reactive crisis management. Hence, the more it will be engaged in clean-up efforts "after the fact". Conversely, the more crises an organization anticipates, the more it will engage in proactive behavior directed towards activities of proactive crisis management.

No matter what, we should realize that nearly every potential crisis is thought to leave a repeated trail of early warning signals. Thus, if organizations could only learn to read these warning signals and to attend to them more effectively, then there is much they could do to prevent many crises from ever occurring.

7.5 Crisis communication

The Chinese character representing the idea of "crisis" means at the same time "danger" and "opportunity": whatever does not kill you makes you stronger, and a crisis, in the end, is nothing but an opportunity to learn from past mistakes and to start over, for an enterprise or an organization, on the condition it keeps its guard up as the next crisis may not be far around the corner. A crucial matter in this respect is the way the crisis is communicated.

To make sure words do not become ills, every term used must be chosen with care and precision by the public powers and the political, economic and media actors. The most challenging part of crisis communication management is reacting – with the right response – quickly. This is because behavior always precedes communication. Non-behavior or inappropriate behavior leads to spin, not communication. In emergencies, it is the non-action and the resulting spin that cause embarrassment, humiliation, prolonged visibility, and unnecessary litigation (11).

Crisis management is primarily a matter of communication. When choosing a single spokesperson, be sure it is someone who speaks well and who knows how to step back and choose his/her register. The main objective is to restore or maintain the trust or confidence. The best communication is therefore not to deny that something is past, but to indicate that everything will be done to understand the facts and take the necessary measures so that the event will not recur. Crisis communication should be prepared and should evolve with the incoming information.

The skills needed for a good communication could be expressed as:

- Keep calm.
- Try to stand back.
- Rapid synthesis and analysis competence.
- Being reactive instead of passive.
- Learn to decide quickly and well.
- Be positive.
- Speaking skills.
- Know how to inform.

Besides, every word counts in a crisis situation. A badly written press release, a defensive or arrogant speech, and the whole communication strategy collapses.

Fuller and Vassie (12) indicate that the effectiveness and efficiency of the communication process depends on the level of attention provided to the communicator by the receiver, the perceptual interpretation of the message by the receiver, the situational

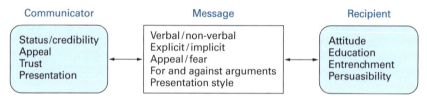

Fig. 7.7: Key elements of the communication process.

context in which the information is provided and the trust that the receiver has in the provider of the information. Key elements of the communication process are summarized in ▶Fig. 7.7.

They also mention that "framing effects", which relate to the context in which information is presented or "framed", can lead to bias in the recipients' views of the risks. The most common framing effect is that of the *domain effect*, which involves changing the description of a risk from a negative to a positive description by, e.g., identifying the benefits associated with the risk rather than the losses. Where a choice must be made between two undesirable options, both options can be framed in terms of their relative gains rather than the losses, so that those people affected by the decision would be left with a positive feeling that they had made a real gain whichever decision was made. Framing effects are routinely employed in the advertising industry and by politicians.

> In crisis situations, the pace of the conflict accelerates dramatically. This means that the parties have to react very quickly to changing conditions or risk as their their ability to protect their interests is substantially reduced. Crises are likely to be further complicated by the increased levels of fear, anger, and hostility that are likely to be present. Often in crises, communication gets distorted or cut off entirely. As a result, rumours often supplant real facts, and worst-case assumptions drive the escalation spiral. In addition, parties often try to keep their real interests, strategies, and tactics secret, and use deceptive strategies to try to increase their relative power. [Inspired from Conflict Research Consortium, University of Colorado, USA (13).]

7.6 Conclusions

When critical services and products cannot be delivered, consequences can be severe. All organizations are at risk and face potential disaster if unprepared. Wherever possible, economically, socially and environmentally sustainable measures must be set up not only to moderate risk, avoid major accidents and crises but also to continuously deliver products and services despite disruption; that will/might happen sooner or later.

References

1. Swartz, E., Elliott, D., Herbane, B. (2003) Greater than the sum of its parts: business continuity management in the UK finance sector. Risk Management 5:65–80.
2. Le Ray, J. (2010) Gérer les risques Pourquoi? Comment? AFNOR. La Plaine Saint-Denis Cedex, France: Afnor Editions.

3. Mitroff, I.I., Pearson, C.M., Harrington, L.K. (1996) The Essential Guide to Managing Corporate Crises. New York: Oxford University Press.
4. Fink, S. (2000) Crisis management – Planning for the Inevitable. Lincoln, NE: iUniverse Inc.
5. Mitroff, I., Pauchant, T., Shrivastava, P. (1988) The structure of man-made organizational crisis. Technol. Forecast Soc. 33:83–107.
6. Hiles, A. (2011) The Definitive Handbook of Business Continuity Management. Chichester, UK: Wiley & Sons.
7. Bhagwati, K. (2006) Managing Safety – A Guide for Executives. Weinheim, Germany: Wiley-VCH.
8. Department of Health & Human Services. USA (accessed in 2012). http://www.hhs.gov/.
9. National Institute of Environmental Health. USA (accessed in 2012). http://www.niehs.nih.gov/.
10. National Oceanic and Atmospheric Administration. USA (accessed in 2012). http://www.noaa.gov/.
11. Ulmer, R.R., Sellnow, T.L., Seeger, M.W. (2010) Effective Crisis Communication: Moving From Crisis to Opportunity. SAGE Publications.
12. Fuller, C.W., Vassie, L.H. (2004) Health and Safety Management. Principles and Best Practice. Essex, UK: Prentice Hall.
13. Conflict Research Consortium, University of Colorado, USA (accessed in 2012). http://conflict.colorado.edu/.

8 Economic issues of safety

It seems evident that risk management and safety management are essential to any manager. However, Perrow (1) indicates that there are indeed reasons why managers and decision-makers would not put safety first. The harm, the consequences, are not evenly distributed: the latency period may be longer than any decision maker's career. Few managers are punished for not putting safety first even after an accident, but will quickly be punished for not putting profits, market share or prestige first. But in the long-term, this approach is obviously not the best management solution for any organization. The economic issues of risks – playing a crucial role in the decision-making on safety management budgets and budget constraints, and also having a profound impact on other organizational budgets such as production budgets, human resources budgets, maintenance budgets, etc. – are explained in this chapter.

One question increasingly on the mind of a lot of corporate senior executives today is: "What are the risk/opportunity trade-offs of investing in safety?" Of course, safety is not a monolith, it might be very heterogeneous. We could draw a parallel with emerging technologies. The rate of growth in emerging technologies for the past decades has been higher than the growth rate of classical technologies. This is why there is increasing interest in emerging technologies by companies in advanced economies, where growth has been much slower. Importantly, these growth differentials reflect a secular transformation in the structure of the global economy, not a cyclical phenomenon occasioned by the current economic/financial crisis.

Accidents and illness at work are matters of health, but they are also matters of economics, as they stem from work, and work is an economic activity. The economic perspective on safety and health encompasses both causes and consequences: the role of economic factors in the search for causes and their effects with respect to the economic prospects for workers, enterprises, nations, and the world as a whole. It is therefore a very broad perspective, but it is not complete, because neither the causation nor the human significance of safety and health can be reduced to its economic elements. Economics means one thing to the specialist and another to the general public.

For economics, a central concept is that of costs. On the one side, we have the costs of improving the conditions of work, in order to reduce the incidence of injury and disease. On the other, we have the costs of not doing these things. Hence, discussion of safety, however defined, involves a discussion of choice. Safety is largely the product of human action consequent upon prior choices. The economics of safety consist in pricing it, or compensating its absence, so as to produce the economically optimal amount of safety at socially optimal cost. Safety may be treated as a resource-absorbing commodity or service. Its production therefore involves the allocation of resources to the production of safety and the allocation of the resulting safety to those in need of it and entitled to it because they have paid for it in some fashion. These two allocative functions can be performed, at least suppositiously, by the market, through other institutional arrangements or by their combination. The object of allocation is to produce safety in the right amount, supply this amount as cost-effectively as possible, and allocate it appropriately. Provision of compensation against the incidence of unsafety is much more suited to regulation by the market than the allocation of safety.

Safety is not costless. Providing it absorbs scarce resources that have alternative uses; they constitute the visible cost of safety. We could always raise the question of: "Should it be worth to invest in stock options the money I intend to invest in safety?"

Again, one answer could be: If you think that safety is expensive .... Try an accident!

Indeed: the fact that safety – or prevention – has a cost does not mean that it does not have a benefit, on the contrary. The next sections will explain the costs and the benefits related to safety.

8.1 Accident costs and hypothetical benefits

Accidents do bear a cost – and often not a small one. An accident can indeed be linked to a variety of direct and indirect costs. ▶Tab. 8.1 below reveals potential socio-economic costs that might accompany accidents.

Hence, by implementing a sound safety policy and by adequately applying engineering risk management, substantial costs can be avoided, namely *all costs related with accidents that have never occurred. We call them "hypothetical benefits"*. However, in reality companies place little or no importance on "hypothetical benefit" because of its complexity.

Non-quantifiable costs are highly dependent on non-generic data such as individual characteristics, a company's culture and/or the company as a whole. Rather, the costs assert themselves when the actual costs supersede the quantifiable costs. In economics (e.g., in environment-related predicaments) monetary evaluation techniques are often used to specify non-quantifiable costs, amongst them the contingent valuation method and the conjoint analysis or hedonic methods (3). In the case of non-quantifiable accident costs, various studies demonstrate that these form a multiple of the quantifiable costs (4–8).

The quantifiable socio-economic accident costs (see ▶Tab. 8.1) can be divided into direct and indirect costs. Direct costs are visible and obvious, while indirect costs are hidden and not immediately evident. In a situation where no accidents occur, the direct costs result in direct hypothetical benefits while the indirect costs result in indirect hypothetical benefits. Resulting indirect hypothetical benefits comprise, e.g., not having sick leave or absence from work, not having staff reductions, not experiencing labor inefficiency, and not experiencing change in the working environment. ▶Fig. 8.1 illustrates this reasoning.

Although hypothetical benefits seem to be rather theoretical and conceptual, they are nonetheless important to fully understand safety economics. Hypothetical benefits can be identified by the enterprise by asking questions such as, "How much does the installation or a part of it cost?" or "What would be the higher price of the premium should a specific accident occur?" etc. Hypothetical benefits resulting from *non-occurring accidents* can be divided into five categories at organizational level:

1. The first category concerns the non-loss of work time and includes the non-payment of the employee who at the time of the accident adds no further value to the company (if payment were to continue this would be a pure cost).

Tab. 8.1: Non-exhaustive list of quantifiable and non-quantifiable socio-economic consequences of accidents [see also (2)].

Interested parties	Non-quantifiable consequences of accidents	Quantifiable consequences of accidents
Victim(s)	• Pain and suffering • Moral and psychic suffering • Loss of physical functioning • Loss of quality of life • Health and domestic problems • Reduced desire to work • Anxiety • Stress	• Loss of salary and bonuses • Limitation of professional skills • Time loss (medical treatment) • Financial loss • Extra costs
Colleagues	• Bad feeling • Anxiety or panic attacks • Reduced desire to work • Anxiety • Stress	• Time loss • Potential loss of bonuses • Heavier work load • Training and guidance of temporary employees
Organization	• Deterioration of social climate • Poor image, bad reputation	• Internal investigation • Transport costs • Medical costs • Lost-time (informing authorities, insurance company, etc.) • Damage to property and material • Reduction in productivity • Reduction in quality • Personnel replacement • New training for staff • Technical interference • Organizational costs • Higher production costs • Higher insurance premiums • Sanctions imposed by parent company • Sanctions imposed by the government • Modernization costs (ventilation, lighting, etc.) after inspection • New accident indirectly caused by accident (due to personnel being tired, inattentive, etc.) • Loss of certification • Loss of customers or suppliers as a direct consequence of the accident • Variety of administrative costs • Loss of bonuses • Loss of interest on lost cash/profits • Loss of shareholder value

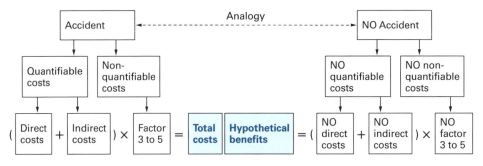

Fig. 8.1: Analogy between total accident costs and hypothetical benefits.

2. The non-loss of short-term assets forms the second category and can include, e.g., the non-loss of raw materials.
3. The third category involves long-term assets, such as the non-loss of machines.
4. Various short-term benefits, such as non-transportation costs and non-fines, constitute the fourth category.
5. The fifth category consists of non-loss of income, non-signature of contracts or non-price reductions.

Clearly the visible or direct hypothetical benefits generated by the avoidance of costs resulting from non-occurring accidents only make up a small portion of the factors responsible for the total hypothetical benefits resulting from non-occurring accidents.

Next to visible benefits, invisible benefits might exist. Invisible benefits may manifest themselves in different ways: e.g., non-deterioration of image, avoidance of lawsuits, the fact that an employee, thanks to the safety policy and the non-occurrence of accidents, does not leave the organization, etc. If an employee does not leave the organization, the most significant benefit arises from work hours that other employees do not have to make up. Management time is consequently not given over to interviews and routines surrounding the termination of a contract. The costs of recruiting new employees can prove considerable if the company has to replace employees with experience and company-specific skills. Research into the reasons that form the basis of recruitment of new staff reveal a difference between, on the one hand, the recruitment of new employees because of the expansion of a company and, on the other hand, the replacement of staff who have resigned due to poor safety policies (7).

> Costs of incidents and accidents that happened ("accident costs") and costs of incidents and accidents that were avoided and that never happened ("hypothetical benefits") are different in their nature. Nonetheless, their analogy is clear and therefore they can easily be confused when making safety cost-benefit considerations.

8.1.1 Quick calculation example of accident costs based on the number of serious accidents

We can use the Bird pyramid from ▶Fig. 3.5 to make a rough estimate of the total yearly costs of occupational incidents and accidents within an organization, based on

8.1 Accident costs and hypothetical benefits | 207

Fig. 8.2: The Bird accident pyramid with costs per type of incident/accident.

the number of serious accidents. Companies usually have a good understanding of the cost per type of accident. For example, the cost of a serious accident is calculated on a yearly basis by organizations, and equals x€. Similarly, the costs of the different types of incidents and accidents from ▶Fig. 3.5, can be determined. ▶Fig. 8.2 shows some theoretical ways of calculating the different costs.

Based on these costs, a rough estimate of the total yearly cost can be made. Legislation requires every company to know the number of serious accidents that happen per year, so let us assume that the number of serious accidents is N, and hence this rough calculation can be done for any company. ▶Tab. 8.2 shows us how to calculate the total yearly accident costs.

▶Tab. 8.2 shows that, based on the ratio between serious accidents and other types of accidents, a rough estimate can be made of the total accident costs for a given year if the number of incidents and accidents in that year is known, and if the cost per type of accident is known.

A calculation example of hypothetical benefits is much more difficult to carry out, as there are no numbers available of the number of serious accidents avoided, or of the avoided cost per type of accident. Future research has to be carried out to gain a better understanding of the hypothetical benefits and their calculation.

Tab. 8.2: Quick calculation of the total yearly costs based on the number of serious accidents.

Type of incident/ accident	Bird pyramid	Number of incidents/ accidents	Cost per type of incident/accident	Cost
Serious	1	N	x	$N.x$
Minor injury	10	$10.N$	y	$10.N.y$
Property damage	30	$30.N$	$z + t$	$30.N.(z + t)$
Incident	600	$600.N$	s	$600.N.s$
			Total cost:	$N.(x + 10.y + 30.(z + t) + 600.s$

8.2 Prevention costs

In order to obtain an overview of the various kinds of prevention costs, it is appropriate to distinguish between fixed and variable prevention costs on the one hand and direct and indirect prevention costs on the other.

Fixed prevention costs remain constant irrespective of changes in a company's activities. One example of a fixed cost is the purchase of a fire-proof door. This is a one-off purchase and the related costs are not subject to variation in accordance with production. Variable costs, in contrast to fixed costs, vary proportionally in accordance with a company's activities. The purchase of safety gloves can be regarded as a variable cost because the gloves have to be replaced sooner when used more frequently or more intensively because of increased productivity levels.

Direct prevention costs have a direct link with production levels. Indirect prevention costs, on the contrary, are not directly linked to production levels. A safety report, e.g., will state where hazardous materials must be stored, but this will not have a direct effect on production. Hence, the development of company safety policy includes not only direct prevention costs such as the application and implementation of safety material, but also indirect prevention costs such as development and training of employees and maintenance of the company safety management system. A non-exhaustive list of prevention costs is given:

- Staffing costs of the company HSE department.
- Staffing costs for the rest of the personnel (time needed to implement safety measures, time required to read working procedures and safety procedures, etc.).
- Procurement and maintenance costs of safety equipment (fire hoses, fire extinguishers, emergency lighting, cardiac defibrillators, pharmacy equipment, etc.).
- Costs related to training and education with regards to working safe.
- Costs related to preventive audits and inspections.
- Costs related to exercises, drills, simulations with regards to safety (e.g., evacuation exercises, etc.).
- A variety of administrative costs.
- Prevention-related costs for early replacements of installation parts, etc.
- Maintenance of machine park, tools, etc.
- Good housekeeping.
- Investigation of near-misses and incidents.

In contrast to quantifying hypothetical benefits, companies are usually very experienced in calculating direct and indirect non-hypothetical costs of preventive measures, as listed above.

8.3 Prevention benefits

An accident that does not occur in an enterprise thanks to the existence of an efficient safety policy within the company, was referred to in Section 8.1 as a *non-occurring accident*. Non-occurring accidents (in other words the prevention of accidents) result in avoidance of a number of costs and thus create hypothetical benefits, as explained before. An estimation of the amount of input and output of the implementation of a safety policy is thus clearly anything but simple. It is impossible to specify the costs and benefits of one exceptional measure. The effectiveness (and the costs and benefits) of a safety policy must be regarded as a whole.

Hence, if an organization is interested in knowing the efficiency and the effectiveness of its safety policy and its prevention investments, in addition to identifying all kinds of prevention costs, it is also worth calculating the hypothetical benefits that result from non-occurring accidents. By taking all prevention costs and all hypothetical benefits into account, the true prevention benefits can be determined.

8.4 The degree of safety and the minimum total cost point

As Fuller and Vassie (8) indicate, the total cost of a company safety policy will be directly related to an organization's standards for health and safety, and thus its degree of safety. The higher the company's H&S standards and its degree of safety, the greater the prevention costs, and the lower the accident costs within the company. Prevention costs will rise exponentially as the degree of safety increases, because of the law of diminishing returns, which describes the difficulties of trying to achieve the last small improvements in performance. If a company has already made huge prevention investments, and the company is thus performing very well on health and safety, it becomes ever more difficult to further improve its H&S performance. On the contrary, the accident costs will decrease exponentially as the degree of safety improves because, as the accident rate is reduced and hence accident costs are decreased, there is ever less potential for further (accident reduction) improvements. ▶Fig. 8.3 illustrates this.

From a purely micro-economic viewpoint, a cost-effective company will chose to establish a degree of safety that allows it to operate at the minimum total cost point (see ▶Fig. 8.3). At this point, the prevention costs are balanced by the accident costs, and at first sight the "optimal safety policy" (from a micro-economic point of view), is realized. It is important that for this exercise to deliver as accurate results as possible, the calculation of both prevention costs and accident costs should be as complete as possible, and

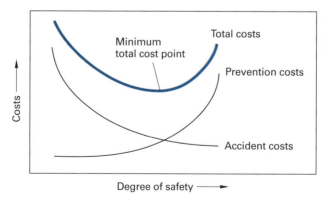

Fig. 8.3: Prevention costs and accident costs as a function of the degree of safety (qualitative figure).

all direct and indirect, visible and invisible, costs should be taken into account. But the exercise results will still be able to be largely improved, and the next paragraph and section explain why and how.

We mention the phrase "from a micro-economic point of view" in the previous paragraph, because victim costs and societal costs should in principle also be taken into account in the total costs calculations. If only micro-economic factors are important for an organization, only accident costs related to the organization are considered by the company, and not the victim costs and the societal costs. The human costs are obvious: loss in quality of life caused by, e.g., pain, stress, incapacity, etc. Financial impacts on society arise from taxes that are used to provide medical services and social security payments for people injured at work, and prices of goods and services that are increased in order to recover the additional operating costs caused by accidents and ill-health at work. Hence, even if all direct and indirect micro-economic accident costs are determined by an organization, there will be an underestimation of the full economic costs, because of individual costs and macro-economic costs.

> True accident costs are composed of the organizational costs, as well as the victim's costs and the society's costs.

Moreover, to really take optimal decisions regarding prevention investments, organizations should make a distinction between the different types of risks (for the different types of risks, see Chapter 2, Section 2.2), and between accident costs and hypothetical benefits.

8.5 Safety economics and the three different types of risks

The optimum degree of safety required to prevent losses is open to question, both from a financial and economic point of view and from a policy point of view. As explained above, developing and executing a sound prevention policy involves prevention costs on the one hand, but the avoidance of accidents and damage leads, on the other hand, to hypothetical benefits. Consequently, in dealing with safety, an organization should try to establish an optimum between prevention costs and hypothetical benefits. Note that the real costs of actual accidents (such as displayed in ▶Fig. 8.3) are not the same costs as those taken into consideration while determining the hypothetical benefits.

It is possible to further expand upon costs and benefits of accidents in general in terms of the degree of safety. The theoretical degree of safety can vary between $(0+\varepsilon)\%$ and $(100-\varepsilon)\%$, wherein ε assumes a (small) value, suggesting that "absolute risk" or "absolute safety/zero risk" in a company is, in reality, not possible. The economic breakeven safety point, namely the point at which the prevention costs equal the resulting hypothetical benefits, can be represented in graph form (see ▶Fig. 8.4). The graph distinguishes between two cases: occupational accidents (or type I accidents where a lot of information is available regarding the accidents) and major accidents (or type II and III accidents where very scarce or no information is available regarding the accidents).

This distinction must be emphasized because there is a considerable difference in the costs and benefits relating to the two different types of accident. In the case of occupational accidents, the hypothetical benefits resulting from non-occurring accidents

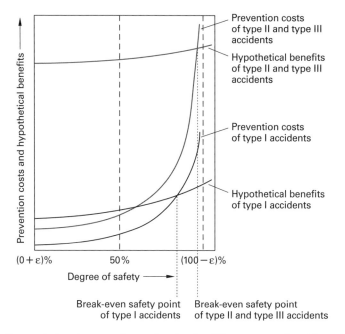

Fig. 8.4: Economic break-even safety points for the different types of accidents (qualitative figure).

are considerably lower than in the case of major accidents. This means essentially that the company reaps greater potential financial benefits from investing in the prevention of major accidents than in the prevention of occupational accidents. When calculated, prevention costs related to major accidents are in general also considerably higher than those to occupational accidents. Major accidents are mostly prevented by means of expensive technical studies (QRAs) and the furnishing of (expensive) technical state-of-the-art equipment within the framework of process safety (e.g., SIL2, SIL3, SIL4). Occupational accidents are therefore more associated with the protection of the individual, first aid, the daily management of safety codes, etc.

As the company invests more in safety, the degree of safety will also increase. Moreover, the higher the degree of safety the more difficult it becomes to improve upon this (i.e., to increase it) and the curve depicting investments in safety thereafter displays asymptotic characteristics. This was also explained in the text accompanying the discussion of ▶Fig. 8.3. Moreover, as more financial resources are invested in safety from the point of $(0+\varepsilon)\%$, higher hypothetical benefits are obtained as a result of non-occurring accidents. These curves will display a much more level trajectory because marginal prevention investments do not produce large additional benefits in non-occurring accidents. It should be stressed that the curves displayed in ▶Fig. 8.4 are qualitative, and that the exact levels at which the curves are drawn, are merely chosen for illustrative purposes, and to increase the understanding of the theory explained for the readers. The hypothetical benefits curve and the prevention costs curve dissect at a break-even safety point. If there are greater prevention costs following this point, hypothetical benefits will no longer balance prevention costs. A different position for the costs and benefits curves is obtained for the different types of accident. ▶Fig. 8.4 illustrates the qualitative benefits curves for the different types of accidents.

It is clear that hypothetical benefits and prevention costs relating to type I accidents are considerably lower than those relating to type II and III accidents. The break-even safety point for type I accidents is likewise lower than that of type II and III accidents. ▶Fig. 8.4 also shows that in the case of type II and III accidents, where a company is subject to extremely high financial damage, the hypothetical benefits are even higher and the break-even safety point must be located near the (100-ε)% degree of safety limit. This supports the argument that in the case of such type of accidents almost all necessary prevention costs can be justified: the hypothetical benefits are certainly (nearly) always higher than the prevention costs. It should be noted that one also has to take into account that the uncertainty levels for type II and III risks, are much higher than those of type I risks, and therefore that the decision to allocate a prevention budget, is not that straightforward.

Thus for type II and III accidents it is possible to state that no real cost-benefit comparison is required but that a very high degree of safety should be guaranteed in spite of the cost of corresponding prevention measures. Of course, the uncertainties associated with these types of accidents are very high, and therefore, managers are not always convinced that such a major accident might happen in their organization. This is the main reason why the necessary precautions are not always taken, and why organizations are not always prepared.

In case of type I risks, regular cost-benefits analyses should be carried out, and the availability of sufficient data should lead to obtaining reliable results, based on which optimal precaution decisions can then be taken.

> Risks possibly leading to occupational accidents (type I) are not to be confused with risks possibly leading to major accidents (type II and III) when making prevention investment decisions. Cost-benefits methods may yield reliable results for type I risks, whereas the hypothetical benefits of type II and III risks almost always outweigh the prevention costs for these types of risks. The uncertainties accompanying the risks should also be taken into account while making decisions.

8.6 Cost-effectiveness analysis and cost-benefit analysis for occupational (type I) accidents

8.6.1 Cost-effectiveness analysis

Cost-effectiveness analysis is employed to maximize the return for a fixed budget. The technique can be used in the safety management field to compare the costs associated with a range of prevention measures and risk control measures that achieve similar benefits, in order to identify the least cost option. The approach can be illustrated by two prevention options, having a different cost and avoiding different numbers of accidents: see ▶Tab. 8.3.

Tab. 8.3: Cost-effectiveness analysis of two prevention options.

	Prevention costs	Number of similar accidents avoided	Average prevention cost per accident avoided
Prevention option 1	€5000	25	200
Prevention option 2	€3240	18	180

8.6 Cost-effectiveness analysis and cost-benefit analysis for occupational (type I) accidents

The average prevention cost per accident avoided is higher for prevention option 1 than for option 2 (see ▶Tab. 8.3), indicating that the second option is more cost-effective than the first. We should remark that to use cost-effectiveness analysis to compare different options, they should share common outputs and be measured in similar terms.

8.6.2 Cost-benefit analysis

Cost-benefit analysis is used to evaluate the ratio between the benefits and the costs of certain measures. Fuller and Vassie (8) indicate that cost-benefit analysis requires financial terms to be assigned to both costs and benefits, to be able to calculate the "cost-benefit ratio". Several approaches are possible for determining such ratios, e.g., the value of averted losses divided by the prevention costs (over their lifetime), or the (liability of the original risk – liability of the residual risk) divided by prevention costs (over their lifetime). If the cost-benefit ratio is greater than 1, the safety management project could be accepted on the grounds that the benefits outweigh the costs, but if the ratio is less than 1 the project can be rejected on the grounds that the costs outweigh the benefits.

▶Tab. 8.4 illustrates the cost-benefit analysis approach, taking the numbers of ▶Tab. 8.3 and adding the assumption that the financial benefit attributable to each accident avoided is €250.

The cost-benefit analysis from ▶Tab. 8.4 shows that prevention option 1 has a cost-benefit ratio of 1.25, whereas option 2 has a ratio of 1.39. The second option is thus clearly preferred; if a cost-benefit analysis would be used.

8.6.2.1 Cost-benefit analysis for safety measures

It is possible to determine in a simple way whether the costs of a safety measure outweighs – or not – its benefits. In general, the idea is simple: compare the cost of the safety measure with its benefits. The cost of a safety measure is easy to determine, but the benefit is much more difficult to calculate. The benefit can be expressed as the "reduced risk", taking into account the costs of accidents with and without the safety measure implementation. The following equation may be used for this exercise (9):

$$\{(C_{without} \cdot F_{without}) - (C_{with} \cdot F_{with})\} \cdot Pr_{control} > \text{Safety measure cost}$$

Tab. 8.4: Cost-benefit analysis of two prevention options.

	Prevention costs	Number of similar accidents avoided	Total benefit	Net benefit	Cost-benefit ratio
Prevention option 1	€5000	25	€6250	€1250+	1.25
Prevention option 2	€3240	18	€4500	€1260+	1.39

Or, if sufficient information is not available for the frequencies of the initiating events to use the previous equation:

$$(C_{without} - C_{with}) \cdot F_{accident} \cdot Pr_{control} > \text{Safety measure cost}$$

With:
$C_{without}$ = cost of accident without safety measure
C_{with} = cost of accident with safety measure
$F_{without}$ = Satiatical frequency of initiating event if the safety measure is not implemented
F_{with} = Satiatical frequency of initiating event if the safety measure is implemented
$F_{accident}$ = statistical frequency of the accident
$Pr_{control}$ = Probability that the safety measure will perform as required

The formulae show immediately why cost-benefit analyses may only be carried out for risks where sufficient data is available: if not, the required "statistical frequencies" are not known, the probabilities may not be known, and rough estimates (more or less *guesses*) should be used, leading to unreliable results. If sufficient information is available, results from using these equations for determining the cost-benefit of a safety measure, are reliable.

8.6.3 Advantages and disadvantages of analyses based on costs and benefits

It should be remembered that the lack of accuracy associated with cost-effectiveness analysis and cost-benefit analyses can give rise to significantly different outcomes in assessments of the same issues by different people. In addition, it is often much easier to assess all kinds of costs, than to identify and to evaluate the benefits.

Cost-benefit analysis should only be used for type I risks, where a sufficient amount of information and data are available to be able to draw sufficiently precise and sufficiently reliable conclusions. If it is used for type II or III risks, it creates an image of accuracy and precision that it does not have. In that case, many of the valuations used for the costs and benefits reflect the perceptions of the person carrying out the analysis rather than their real values. More research is needed to develop reliable cost-benefit analyses or cost-effectiveness analyses for type II or III risks.

Moreover, it is often difficult to incorporate realistic calculations of the net present value of future costs and benefits into the analyses. The net present value (NPV) can be defined as the net value on a given date of a payment or series of payments made at other times. If the payments are made in the future, they are discounted to reflect the time value of money and other factors such as investment risk. Net present value calculations are widely used in business and economics to provide a means of comparing cash flows at different times on a meaningful "like by like" basis. Discounted values reflect the reality that a sum of money is worth more today than the same sum of money at some time in the future. Therefore, prevention costs incurred today should be compared with hypothetical benefits obtained at some time in the future, but equated to today's values. To achieve a benefit equal to €B in N years' time, one must have a benefit with a current net present value of €B/(1 + interest rate)^N.

Furthermore, Frick (10) claims that the cost-benefit approach underestimates the benefits and overestimates the costs of health and safety improvement programmes.

Techniques based on costs and benefits that are used to make prevention decisions have one major undeniable strength: if used correctly, they allow us to allocate limited financial resources efficiently for occupational (type I) risks, and, if used with much caution, they provide us with some background knowledge for allocating financial resources aimed to deal with major accident (type II and III) risks.

8.7 Optimal allocation strategy for the safety budget

Safety measures show a diminishing marginal rate of return on investment: further increases in the number of a type of safety measure become ever less cost effective, and the improvements in safety benefits per extra safety measure exhibit a decreasing marginal development (8). In other words, the first safety measure of type "technology" provides the most safety benefit, the second safety measure provides less safety benefit than the first, and so on. Hence, there should be safety measures chosen from different types (procedures, people, and technology) to be most efficient. Of course, there may be differences in the safety benefit curves for the different types of safety measures.

▶Fig. 8.5 shows the increased safety benefits from choosing a variety of safety measures.

▶Fig. 8.5 shows that, if the total budget available is spread over a range of procedures-related, people-related, and technology-related safety measures, the overall safety benefit can be raised from point A (only investment in procedures-related safety measures) to point B (investment in P, P, T – related safety measures).

> Spreading the safety budget over different types of safety measures is always more efficient and effective than only focusing on one type of safety measure.

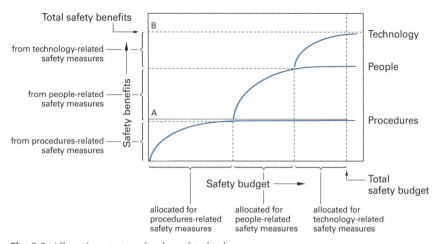

Fig. 8.5: Allocation strategy for the safety budget.

8.8 Loss aversion and safety investments – safety as economic value

Because of the psychological principle of "loss aversion" (11) and the fact that people hate to lose, safety investments to manage and control all types of accidents, but especially precautionary investments to deal with highly unlikely events, are not at all evident. Risk managers, being human beings like all other people, also may let their decision judgment be influenced by this psychological principle.

To have a clear idea of "loss aversion", the following example can be given. Suppose you are offered two options: (A) You receive €5000 from me (with certainty) or (B) We toss a coin and you receive €10,000 from me if it is heads, but if it is tails you receive nothing.

What will you chose? By far most of the people will choose option (A). They go for the certainty, and prefer €5000 for certain than to gamble and to have nothing.

Let's now consider two different options: (C) You have to pay me €5000 (with certainty); or (D) We toss a coin and you need to pay me €10,000 if the coin turns up heads, but in case of tails you do not need to pay me anything.

What option will you prefer this time? By far most people in this case will prefer option (D). Hence, they go for taking the gamble, and risking paying €10,000 with a level of uncertainty (there is a 50% probability that they will not have to pay anything) instead of paying €5000 for certain.

From this example, it is clear that people hate to lose and that they love certain gains. People are more inclined to take risks to avoid certain losses than they are inclined to take risks to gain uncertain gains.

Translating this psychological principle into safety terminology, it is clear that company management would be more inclined to invest in production ("certain gains") than to invest in prevention ("uncertain gains"). Also, management is more inclined to risk highly improbable accidents ("uncertain losses") than to make large investments ("certain losses") in dealing with such accidents.

Therefore, management should be well aware of this basic psychological principle, and when making prevention investment decisions they should take this into account. The fact that we as human beings are prejudiced, and that we have some predetermined preferences in our minds, should thus really be considered in the decision-making process of risk managers.

Furthermore, safety and accident risk are not adequately incorporated into the economic planning and decision processes. What are the business incentives for investing into safety? There is a need for demonstrating that safety measures have a value in an economic sense. To what extent is it true that businesses would not invest in higher safety if such values cannot be demonstrated?

An overinvestment in safety measures is very likely if we ignore the fact that there is access to an insurance market, while an underinvestment in safety measures is very likely if we purchase insurance without paying attention to that the probability and consequences can be reduced by safety measures.

Abrhamsen and Asche (12) stated that the final decision with respect to how much resource should be spent on safety measures and insurance may be very different dependent on what type of risks that are considered. It makes a difference if it is risks we voluntarily assume for ourself, or if it is a risk imposed by others. Clearly there is more reason for society to enforce standards in the latter case. However, the decision-criterion itself is independent on the type of risk: an expected utility maximizer should combine insurance, invest in safety measures and directly take the costs of an accident, such that the marginal utility of the different actions are the same.

> The fact that decision-makers have an in-built psychological preference to avoid losses, should be consciously considered by these decision-makers when making precaution investment decisions.

8.9 Conclusions

Economic issues of safety constitute much more than calculating the costs of accidents, or determining the costs of prevention. Hypothetical benefits, the benefits gained from accidents that have never occurred, should be considered, and the three types of risk should be taken into account when dealing with prevention investment choices. Decisions concerning safety investments make up for a complex decision problem where opportunity costs, perception and human psychology, budget allocation strategies, the choice of cost-benefits methodologies, etc. all play an important role.

References

1. Perrow, Ch. (2006) The limits of safety: the enhancement of a theory of accidents. In: Key Readings in Crisis Management. Systems and Structure for Prevention and Recovery. Smith & Elliott, editors. Abingdon, UK: Routledge.
2. Reniers, G., Audenaert, A. (2009) Chemical plant innovative safety investments decision-support methodology. J. Saf. Res. 40:411–419.
3. Hanley, N., Spash, C.L. (1993) Cost-Benefit Analysis and the Environment. Cheltenham, UK: Edward Elgar Publishing.
4. Mossink, J., Licher, F. (1997) Costs and Benefits of Occupational Safety and Health, Proceedings of the European Conference on Costs and Benefits of Occupational Safety and Health 1997, The Hague, The Netherlands.
5. Rikhardson, P.M., Impgaard, M. (2004) Corporate cost of occupational accidents: an activity-based analysis. Accid. Anal. Prevent. 36:173–182.
6. Leigh, J.P., Waehrer, G., Miller, T.R., Keenan, C. (2004) Costs of occupational injury and illnesses across industries. Scand. J. Work Environ. Health 30:199–205.
7. Chartered Institute of Personnel and Development (2005) Recruitment, Retention and Turnover. London, UK: CIPD.
8. Fuller, C.W., Vassie, L.H. (2004) Health and Safety Management. Principles and Best Practice, Essex, UK: Prentice Hall.
9. International Association of Oil and gas Producers (2000) Fire System Integrity Assurance, Report No. 6.85/304. London, UK: OGP.
10. Frick, K. (2000) Uses and abuses. In: The Role of CBA in Decision-making, pp. 12–13. Bilbao, Spain: European Agency for Safety and Health at Work.
11. Tversky, A., Kahneman, D. (2004) Loss Aversion in Riskless Choice: A Reference-Dependent Model. In: Preference, Belief, and Similarity: Selected Writings. Shafir. Editor. Cambridge, MA: MIT Press.
12. Abrahamsen, E.B., Asche, F. (2010) The insurance market's influence on investments in safety measures. Safety Sci. (48) 1279–1285.

9 Risk governance

Most of us are taught to think about the long-term consequences of our actions, but it is a life lesson that is easily forgotten – both on an individual and an organizational level. This is why, each year, the World Economic Forum poses the question, "What risks should the world's leaders be addressing over the next 10 years?"

9.1 Introduction

We all know the phrase "It's the economy, stupid!" which James Carville had coined as a campaign strategist of Bill Clinton's successful 1992 presidential campaign. We can easily adapt this phrase for any organization going bankrupt, having lots of losses or having major financial problems, into "It's the risk governance, stupid!" *Risk governance can be defined* (1) *as the totality of actors, rules, conventions, processes and mechanisms concerned with how relevant risk information is collected, analyzed and communicated, and how management decisions are taken.*

Renn (1) provides several reasons why risk governance is crucial in today's organizations. Firstly, risk plays a major role in our society: Ulrich Beck, the famous social scientist, called his book on reflexive modernity *The Risk Society*. In his book, Beck argues that risk is an essential part of modern society, and hence, also the governance of risks. People, including customers, politicians and regulators, thus expect organizations to adequately govern their risks. Secondly, risk has a direct impact upon our life. People, especially employees of organizations, die, suffer, get ill, or experience minor and major losses because they have ignored or misjudged risks, miscalculated the uncertainties or had too much confidence in their ability to master dangerous situations. Governing risks in a solid way, by everyone being part of an organization, would be a solution to this problem. Thirdly, risk is a truly interdisciplinary phenomenon: risk, or domains or factors of it, are studied in natural, medical, engineering, social, cultural, psychological, legal, and other disciplines. None of these science disciplines are able to grasp the holistic substance of risk; only if they combine forces, is a truly adequate approach to understanding and managing risks possible. Risk governance can be employed to ensure this. Fourthly, risk is a concept that links the professional with the private person. If someone gets more experienced in dealing with risks in his professional life, in whatever way, they will also make better decisions in their personal life. An understanding of risk and risk governance directly contributes to this fact.

Risk governance starts with good corporate governance and integrated board management. Hilb (2) indicates that integrated board management has four preconditions for being successful in developing, implementing, and controlling fortunate organizations:

1. Diversity: strategically targeted composition of the board team.
2. Trust: constructive and open-minded board culture.

3. Network: efficient board structure.
4. Vision: stakeholder-oriented board measures of success.

Although these preconditions of success have been proven in a variety of studies, they seem to be very hard to achieve by organizations. In the light of the recent economic crises, Peter Senge asks: "How can a team of committed board members with individual IQs above 120 have a collective IQ of 60?" (2).

As Fuller and Vassie (3) explain, one view of business is that the directors of a company are merely agents acting on behalf of the shareholders and, as such, their sole responsibility is to maximize the return on the investments of these owners: this is referred to as the "principal agent theory". This shareholder-model is referred to as the "Anglo Saxon model". Corporate social responsibility, on the other hand, is a concept derived from a wider perspective that businesses have responsibilities to a range of people in addition to shareholders: this is referred to as the "stakeholder theory", and the stakeholder-model is referred to as the "Rijnland-model".

Whether applying one model or the other for corporate governance, adequate risk governance, as part of good corporate governance, is absolutely necessary for any organization to be healthy in the long-term.

As stated by IRGC (International Risk Governance Council) in (4), "risk governance" involves the "translation" of the substance and core principles of governance to the context of risk and risk-related decision-making. Risk governance includes the totality of actors, rules, conventions, processes, and mechanisms concerned with how relevant risk information is collected, analyzed and communicated and management decisions are taken.

Encompassing the combined risk-relevant decisions and actions of both governmental and private actors, risk governance is of particular importance in, but not restricted to, situations where there is no single authority to take a binding risk management decision but where, instead, the nature of the risk requires the collaboration of, and coordination between, a range of different stakeholders. Risk governance, however, not only includes a multifaceted, multi-actor risk process, but also calls for the consideration of contextual factors, such as institutional arrangements (e.g., the regulatory and legal framework that determines the relationship, roles and responsibilities of the actors and coordination mechanisms such as markets, incentives or self-imposed norms) and political culture, including different perceptions of risk.

Risk governance therefore requires dealing with risks and uncertainties in a very holistic and general way, with the necessary procedures in place at high, strategic levels, besides the obvious operational and tactic levels, at different levels of the organization, and even inter-organizational, with collaboration as much as needed but as low as required (to avoid complexity), with a diversity of people, experts and disciplines, and above all with an open mind. We will discuss a system and its requirements (Section 9.2), a framework (Section 9.3), and a model (Section 9.4) that together lead to effective risk governance, on an operational, tactic and strategic level.

9.2 Risk management system

Management systems for the safe operation of organizations require a system of structures, responsibilities, procedures, and the availability of appropriate resources and

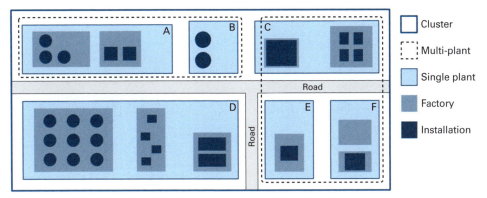

Fig. 9.1: Comparisons of installation, factory, plant, multi-plant and cluster [inspired by (5)].

technological know-how. Risks can be managed at different levels: at factory, at plant, at multi-plant and at cluster level. ▶Fig. 9.1 illustrates the difference between these terms, and differentiates between the different levels. Plant B e.g., consists of two installations within a single factory. Furthermore, Plant B belongs to the multi-plant area consisting of plants A and B, and is part of the larger cluster (or, in other words, industrial area) composed of plants A, B, C, D, E, and F.

Factory-level risk management includes topics such as working procedures, work packages, installation-specific training, personal protective equipment, quality inspection, etc. Plant-level risk management includes defining acceptable risk levels and documenting plant-specific guidelines for implementing and achieving these levels for every facility situated on the premises of the plant. It is current industrial practice to draft an organizational *safety management system (SMS)* to meet these goals (see further in this section). Multi-plant or cluster level risk management topics include defining risk level standards by different plants in collaboration, defining risk management cooperation levels, defining acceptable risks involving more than one plant, joint workforce planning and joint emergency planning in the event of cross-plant accidents, etc. The latter multiple-plant related topics can be documented in a type of cross-plant SMS dealing with these issues.

In the nuclear industry or in the chemical industry, e.g., at factory level and at plant level, safety documents, guidelines and instructions, technical as well as meta-technical, are usually very well elaborated. An adapted version of the *plan-do-check-act* process is often used for continuously improving risk management efforts and safety (see also Chapter 3). To optimize risk management, the circular continuous improvement process can be used at all levels within the industry: from top (the cluster level) to bottom (the installation level). To be able to perform the process at all levels, a risk management cycle structure can be established at each level and provided with communication and cooperation links between the different levels. These links are necessary for further optimization of the different levels of looping risk management and for the prevention of double elaboration of certain multi-levelled risk management topics. Such a framework characterized by loop level risk management can be arranged as illustrated in ▶Fig. 9.2.

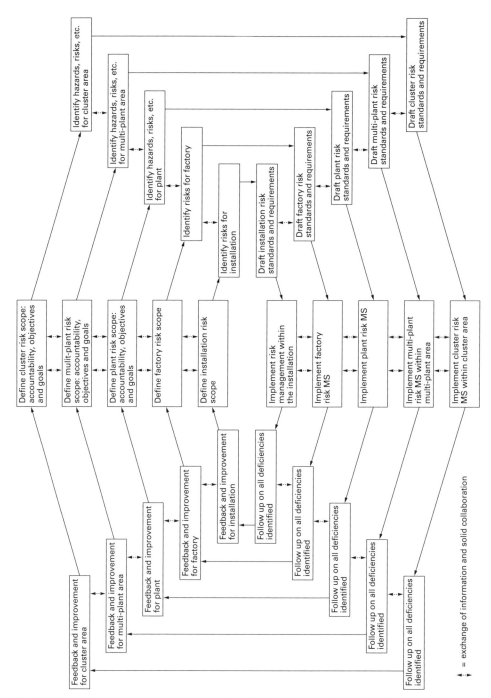

Fig. 9.2: Loop risk management structure on installation, factory, plant, multi-plant, and cluster level [inspired from (5)].

Many organizations already follow the plan-do-check-act loop because of their acquired know-how of internationally accepted business standards, e.g., ISO 9001, ISO 14001, or/and OHSAS 18001 (see also Chapter 1), addressing quality, environmental, and safety management respectively, and continuously improving performance concerning those related risks. Hence, some degree of basic standardization for operational risk governance already exists in many organizations and thorough documented and well-implemented risk management systems are available. One of the areas that is very important in the context of this book, is safety management.

A safety management system (SMS), as part of the risk management system, aims to ensure the various safety risks posed by operating the facility are always below predefined and generally accepted company safety risk levels. Effective management procedures adopt a systematic and proactive approach to the evaluation and management of the plant, its products and its human resources.

To enhance safety for type II and III risks, the SMS considers safety features throughout scenario selection and process selection, inherent safety and process design, industrial activity realization, commissioning, beneficial production and decommissioning. To enhance safety-related to type I risks, both personal and group safety equipment is provided, training programs are installed, and task capabilities are checked. Arrangements are made to guarantee that the means provided for safe operation of the industrial activity are properly designed, constructed, tested, operated, inspected and maintained and that persons working on the site (contractors included) are properly instructed.

Four indispensable features for establishing an organizational SMS are:

- The parties involved.
- The policy – objectives.
- The list of actions to be taken.
- Implementation of the system.

The essence of accident prevention practices consists of safety data, hazard reviews, operating procedures and training. These elements need to be integrated into a safety management document that is implemented in the organization on an on-going basis. To enhance implementation efficiency, this can be divided into 12 subjects (6):

1. Safe work practices – A system should be installed to guarantee that safe work practices are carried out in an organization through procedural and administrative control of work activities, critical operating steps and critical parameters, through pre-startup safety reviews for new and modified plant equipment and facilities, and through management of change procedures for plant equipment and processes.
2. Safety Training – The necessity to periodically organize training sessions emerges from the continuously changing environment of plants, installations, and installation equipment. Employees and contractors at all levels should be equipped with the knowledge, skills and attitudes relating to the operation or maintenance of organizational tasks and processes, so as to work in a safe and reliable manner. Safety training sessions should also lead to a more efficient handling of any incident or accident.
3. Group meetings – An organization should establish a safety group meeting for the purpose of improving, promoting and reviewing of all matters related to safety and health of employees. This way, communication and cooperation between

management, employees and contractors is promoted, ensuring that safety issues are addressed and appropriate actions are taken to achieve and maintain a safe working environment.
4. Pursuing in-house safety rules and complying with regulations.
5. A set of basic safety rules and regulations should be formulated in the organization to regulate safety and health behaviors. The rules and regulations should be documented and effectively communicated to all employees and contractors through promotion, training, or other means, and should be made readily available to all employees and contractors. They should be effectively implemented and enforced within the organization. The company rules should be in conformance with the legislative requirements and rules that are non-statutory should conform to international standards and best practices.
6. Safety promotion – Promotional programs should be developed and conducted to demonstrate the organization's management commitment and leadership in promoting good safety and health behaviors and practices.
7. Contractor and employee evaluation, selection and control – The organization should establish and document a system for assessment and evaluation of contractors to guarantee that only competent and qualified contractors are selected and permitted to carry out contracted works. This way, personnel under external management, but working within the organization, are treated, evaluated and rewarded in the same manner (concerning safety issues) as internally managed personnel.
8. Safety inspection, monitoring and auditing – The organization needs to develop and implement a written program for formal and planned safety inspections to be carried out. The program should include safety inspections, plant and equipment inspections, any other inspections (including surprise inspections), and safety auditing. This way, a system is established to verify compliance with the relevant regulatory requirements, in-house safety rules and regulations and safe work practices.
9. Maintenance regimes – A maintenance program needs to be established to ensure the mechanical integrity of critical plant equipment. In fact, all machinery and equipment used in the organization needs to be maintained at all times so as to prevent any failure of this equipment and to avoid unsafe situations.
10. Hazard analysis and incident investigation and analysis – All hazards in the organization need to be methodically identified, evaluated and controlled. The process of hazard analysis should be thoroughly documented. Written procedures should also be established to ensure that all incidents and accidents (including those by contractors) are reported and recorded properly. Furthermore, procedures for incident and accident investigation and analysis so as to identify root causes and to implement effective corrective measures or systems to prevent recurrence, should be installed.
11. Control of movement and use of dangerous goods – A system should be established to identify and manage all dangerous goods through the provision of material safety data sheets and procedures for the proper use, storage, handling, and movement of hazardous chemicals. To further ensure that all up-to-date information on the storage, use, handling and movement of dangerous goods in the organization reaches the prevention and risk management department, a continuously adjusted database with information should be established.
12. Documentation control and records – An organization should establish a central documentation control and record system to integrate all documentation requirements and to ensure that they are complied with.

9.2 Risk management system

These recommendations can be generalized to multi-plant or cluster-related recommendations [please see (5)].

When drafting the standards and requirements, the Health and Safety Executive (7) indicates that it is essential to ensure standards are incorporated into business input activities through the following [see also (3)]:

- Employees – such as defining physical, intellectual and mental abilities through job specifications based on risk assessment.
- Design and selection of premises – such as consideration of the proposed and foreseeable uses, construction and contract specification.
- Design and selection of plant – such as installation, operation, maintenance, and decommissioning.
- Use of hazardous substances – such as the incorporation of the principle of inherent safety and the selection of competent suppliers.
- Use of contractors – such as selection procedures.
- Acquisitions and divestitures – such as the identification of short-term and long-term safety risks associated with the organization's activities.
- Information – such as maintaining an up-to-date system for relevant health and safety legislation, standards and codes of practice.

The Health and Safety Executive (7) [see also (3)] further lists some factors that should be included in an assessment of workplace safety standards:

- Safety management system – such as policy, organization, implementation, monitoring, audit and review.
- Use of hazardous goods – such as the receipt, storage, use, and transportation of chemicals.
- Use of contractors – such as the provision of working documents and performance reviews.
- Emergency planning – such as the identification of emergency scenarios, liaison with the emergency services and the implementation of emergency planning exercises.
- Disaster and contingency planning – such as the identification of disaster scenarios, preparation of contingency plans, and the implementation of disaster planning exercises.

> Risk management systems are a must for organizations to handle risks at an operational level. Risks are diversified and omnipresent. For example, the chance that I will never finish typing this text exists, although its likelihood is rather very low. Several risks exist in this regard: the chair I am sitting in might break down, and I may fall, thereby crashing the laptop, or I may strike my head. My office may be set on fire. The ceiling may collapse upon me, or my laptop. A plane might crash down on the building I'm sitting in. A bomb may explode very nearby, etc. Risk management systems deal with assessing all these risks (estimating their consequences and likelihoods) and many others, and with treating them, that is, trying to prevent them, and, in the case of an unfortunate event happening despite all measures taken, in trying to mitigate the consequences, or transferring them, etc.

9.3 A framework for risk and uncertainty governance

On the surface, risk management in an organization seems to be all about avoiding any type of unwanted event. All undesired happenings, either with large consequences or with a minor outcome, or with high or low likelihood, are treated by risk management within a company on the same psychological level. But as we have argued before in this book, this psychological level, and thus the management decisions taken, should be different for different types of risks.

It is obvious that, taking the positive side of risks into account, a decision-maker can be risk taking in the case where a lot of historical data is available, as he has some knowledge from past events in this area and the possible negative outcomes (which are never extremely severe in this case) can be predicted quite accurately by using scientific models, statistical methods, etc., because sufficient information is available. Hence, making profits in this type I area follows from taking positive risks and keeping the negative sides of these risks well under control. Profits are tangible and follow from investment choices, production decisions and risk management strategies (amongst others).

However, a decision-maker must be very careful to be risk taking (for making profits) in the area of type II and III risks and uncertainties, as the possible negative consequences in these areas are possibly very high. Actually, making "profits" in the type II and type III areas follows from averting risks: profits in these areas are intangible and hypothetical. Non-occurring accidents (and their accompanying costs) (see also Chapter 8) resulting from risk averting behavior of a decision-maker, should be regarded as a true and large hypothetical benefit in these areas. In addition to classic industrial sectors with the possibility of major accidents such as the chemical industry or the nuclear industry, the banking sector can also be used as an example. We should, e.g., be cautious taking risks in an economic growth period for making huge profits (which is possible in the type II and III areas), as in the event that a worldwide economic crisis occurs (as was the case in 2008), financial disaster may strike. Hence, non-occurring huge financial losses in a bank (which might be realized in the case of a sudden global economic crisis) as the result of careful risk management (aimed at making profits in the type I area) in the economic prosperous period, should be regarded as true and huge hypothetical gains in the type II and III areas.

In summary, decisions to take risks, or indeed to avert them, depend on the character of the risks (i.e. the type of the risks) and their accompanying uncertainties and objectives. Moreover, hypothetical (short-term and long-term) benefits should be taken into account when taking decisions concerning type II and III uncertainties and risks.

A framework for risk and uncertainty governance is suggested in ▶Fig. 9.3 to meet with all the aforementioned goals.

The framework from ▶Fig. 9.3 first considers the different types of risk depending on the amount of data and information that is available [see also (8)]. For the different types of risks, different sets of stakeholders are involved. The more uncertainty there is about the risk, the more stakeholders need to be involved, as not only is the know-how to deal with risk assessment techniques required, but also open-mindedness, the willingness to collaborate, and viewpoints, know-how and knowledge from other, non-purely technological disciplines, which can make the difference to make good or poor risk decisions. In the next step of the framework, risk assessments are carried out. A risk assessment method that can, e.g., be used for all three types of risks, but especially for type III risks, is scenario building. Scenario building is well-known and much used by risk

9.3 A framework for risk and uncertainty governance | 227

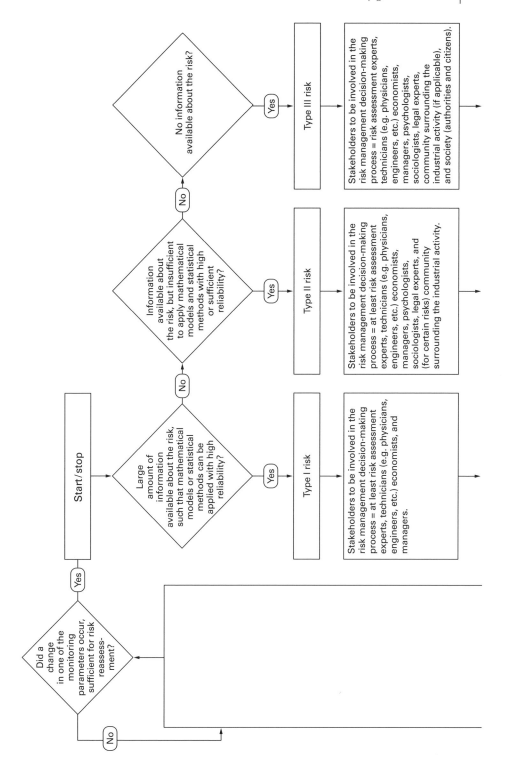

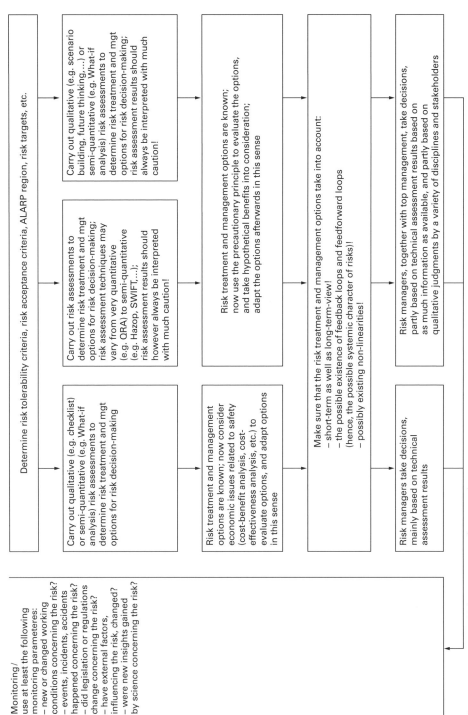

Fig. 9.3: Uncertainty/risk governance framework.

experts. Scenarios are drafted based on past events or by imagining future events using a variety of available techniques, such as classic risk analysis techniques, the Delphi method, etc. Scenarios lead to a better understanding of possible futures, and what can be done to prevent some of the possible unwanted futures, and/or to enhance the emergence of possible wanted futures. It should be noted that the more information is available in case of type II and/or III risks, the less uncertainty there is, and the more quantitative character the used risk assessment technique can have when dealing with these type II and/or III risks. With no or extremely little data or information available, highly quantitative analyses should be avoided. At the best, semi-quantitative analyses might be employed for extremely high uncertainties (type III events), and the results should always be treated and interpreted with very much caution.

When the decision options are known, some further principles should be followed when taking decisions: and at least the precautionary principle. The precautionary principle (9) states that it is not because a cause-consequence relationship between any two variables cannot be proven, that such a relationship does not exist; in other words, unknown complicated relationships (characterizing any complex phenomenon) should not be disregarded, whether the consequences are desired or undesired.

In the last step before carrying out the decision, a decision-maker should reflect on several important systemic principles. It is therefore recommended that every decision is approached by thinking in the long-term (as well as in the short-term), thinking circular (as well as cause-consequence-minded), and thinking non-linear (as well as linear) regarding potential results. Long-term oriented thinking indicates that risk decisions should not only be taking short- and medium-term into account, but also the (very) long-term, thus leading to more sustainable risk decisions. Circular thinking and non-linear thinking are concepts used in systems thinking (10,11). As mentioned before, system thinkers see wholes that function rather than input-output transformers; they also see the parts of the whole and the relationships among them; they see patterns of change rather than static snapshots. Positive and negative feedback loops exist between events. These feedbacks are not necessarily linear, meaning that one should think in terms of "changes of A and B in the same or opposite direction", rather than "a predefined increase or decrease of B caused by a specific increase or decrease of A". Such an approach guarantees more profound and well-considered decisions.

When decisions have been taken, the decision-maker should regularly monitor to check whether no changes have occurred concerning the decision. If this was the case, the risk should be subject to the risk governance framework again.

Summarizing, this risk governance framework aims to obtain objective and consistent results, for the different existing types of risks and uncertainties. Assessments that are subject to small uncertainties should be treated differently from assessments subject to large uncertainties, in a way that under different assumptions and/or the use of other analysts, the decision options and – results concerning any uncertainty or risk – should not be different. Solutions and decisions should only depend – or partially depend – on probabilities and probability estimates under certain conditions and circumstances. As Aven (12) already suggests, risk assessments need to provide a much broader risk picture than is typically the case today. Separate "uncertainty analyses" should be carried out, extending the traditional probability-based analyses. The uncertainties accompanying the risks should be mentioned with every risk assessment result, and they should be used by the risk decision-makers. Only this way, more objective and higher-quality risk decisions can be made. *Risk managers should be uncertainty experts!*

The framework is generic in its application, indicating that any uncertainty/risk can be tackled by it, and recognizing the negative as well as positive consequences of risks. In general terms, the negative side of risks should always be minimized and the positive side should always be maximized. This framework recognizes that this is not always the best solution, leading to optimal decisions, and addresses the need to sometimes "balance" negative and positive consequences rather than opt for one or the other, to be able to ensure long-term success of – and long-term profits within – any organization. By continuously monitoring and questioning its own decisions, the risk governance framework measures up to the science-paradigm of always being critical and questioning any findings.

Company management should be aware that uncertainties and risks should not be rooted out, and that they should not be considered as some kind of evil that detracts from managers' abilities to manage with control in an organization. Uncertainty – and thus risk – should be recognized by company decision-makers as two-sided, creating obstacles for the organization in generating profits and ensuring consistent performance, as well as presenting opportunities for improvement and innovation. Moreover, different levels of uncertainty require different stakeholders, different levels of quantitativeness of risk assessment techniques, different principles to be followed, etc. The risk governance framework takes this into account and offers decision-makers within organizations the possibility not only to manage (negative) risks, but to manage uncertainties. Such an approach leads to better risk decision-making and long-term organizational success.

> The perception of the situation, the circumstances, reality, etc., is essential when assessing risks. Human perception is based on recognition and interpretation of patterns. All perceptions are therefore colored and influenced by experience and expertise, which – intuitively – anticipates everything we encounter, and makes an interpretation. This (scientifically called) "bias", taking the form of expectations, tendencies, and premonitions, helped us to survive in the past. Indeed, making decisions based on very limited information was absolutely necessary (and a major advantage) to be able to survive. However, we do not live in a simple world anymore, dealing with simple tasks and simple processes, where only simple accidents could happen that led to minor injuries or to single deaths at most. We are currently part of a very complex global society with complicated and complex industrial activities. The activities carried out in today's organizations may wipe out entire villages and even cities. Moreover, many people are connected via internet, social media, etc., and hence an incident and its potential consequences, or an accident and its real consequences, are swiftly shared by millions of people at the same time. This completely changed industrial environment requires a whole different way of dealing with risks compared with any time more than 50 years ago.
>
> It is indeed obvious: since about 50 years ago the time of simplicity and of simple single brain heuristics is definitely over. The post-atomic age requires us to find new ways of dealing with all the existing complexities and making the right decisions despite these complexities. The answer to this, in the form of collaboration and perception-improvement (often against intuition), is risk governance, and the risk governance framework helps to attain to the right decisions.

9.4 The risk governance model (RGM)

The most influential research into how people manage risk and uncertainty has been conducted by Kahneman and Tversky, two psychologists (13). One of their most interesting findings is the asymmetry between the way people make decisions involving gains and decisions involving losses. This topic is also mentioned in Chapters 8 and 11, and explained in the light of what is discussed in each. The research results show that when the choice involves losses, people are risk-seekers, whereas when the choice involves gains, people are risk averse. In other words, people tend to not gamble with certain gains, but they tend to gamble with uncertain losses. Other research results indicate inconsistency with the assumptions of rational behavior and people being loss-averse, rather than risk averse. People are apparently perfectly willing to choose a gamble when they consider it appropriate, and they not so much hate uncertainty but they hate losing. Losses are likely to provoke intense, irrational, and abiding risk-aversion, because people are much more sensitive to negative than to positive stimuli.

These results from human psychology and human decision-making have important repercussions on risk decision-making in companies, especially concerning negative risks. Operational risk managers and middle or top management tend to gamble with uncertain losses, possibly leading to major accidents. They often fail to recognize the huge hypothetical benefits resulting from prevention.

Hence, because of the inconsistent understanding and meaning of the risk concept by people and the irrational risk decision-making, it is essential that "risks" should be regarded as "uncertainties" in companies. The less uncertainty there is concerning the possible consequences of a decision, the more it is possible to make adequate and good decisions. This is typically the case for type I risks, and much less the case for type II or type III risks.

But is there a model that can be used for risk governance? Indeed there is. Reniers (14) indicates that the use of a number of triangles allows any organization to optimize its risk decision and expertise process. In the model, 12 triangles are employed (see ▶Fig. 9.4). Note that, to further generalize the model, a thirteenth triangle might be added, on the same level as the "risk" triangle, and describing positive risk: opportunities-exposure-gains. We did not include this triangle into the figure, in order to keep the focus on negative risks, which are the main concern of this book.

Following the well-known PDCA loop of continuous improvement from quality management, a four-step plan (policy-decision-risk-culture or the PDRC loop of continuous improvement) is proposed as the basic structure to serve for the RGM. From a holistic viewpoint, risks, and their uncertainties, outcomes and management, should be the concern of organizations-authorities-academia, these three actors within society forming the first triangle – "risk policy". This first triangle should be the cornerstone of solid and holistic risk management, creating the right circumstances and helping to induce collaboration between all parties involved. The second triangle, "decision", – consists of information-options-preferences. It is obvious that decision-making always requires information, options and preferences as without any one of these three, decisions can simply not be made. Each of the blocks of the subsequent triangle of the rad needs to be aware of the three blocks of the previous triangle of the rad: risk information has to be taken from academia, organizations and authorities, and the same holds for developing options and mapping or composing preferences. The third triangle, "risk", includes hazards, exposure and losses. For the different dimensions of the risk triangle, the decision triangle has to be considered. The fourth triangle, "organizational culture", includes people-procedures-technology. Each

232 — 9 Risk governance

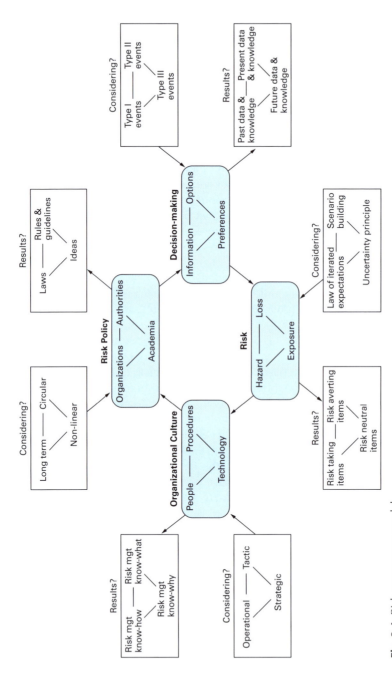

Fig. 9.4: Risk governance model.

of these three domains composing an organizational culture has to take the risk triangle into consideration. In its turn, the organizational culture triangle serves as a guide for each of the domains within the risk policy triangle.

Hence, risk policy guidelines, rules, etc., made or investigated by authorities, organizations, and academia, should ultimately lead to an efficient and effective organizational risk culture. This, in turn, should be used by risk policy makers, as an input for risk policy guidelines, rules, etc., to continuously improve risk decision-making, and so on. People, procedures and technology, forming the backbone of this culture, should be continuously optimized through the PDRC loop displayed in ▶Fig. 9.4.

An effective and efficient organizational culture implies that organizations are open-minded towards collaborating with other organizations and that they are prepared to search for the optimal way to decrease risks, also multi-plant risks, and act accordingly (e.g., in their investment policies).

9.4.1 The "considering?" layer of the risk governance model

"Considering?" triangles can be used to have a holistic indication of how to deal with the PDRC building blocks, and what features should be taken into account when addressing the blocks belonging to the rad of the RGM. For example, it is recommended that academia-organizations-authorities within the risk policy triangle are all approached by thinking in the long-term, thinking circular, and thinking non-linear. Hence, as far as risks and risk decisions and expertise are concerned, the way to think about people, procedures and technology by academia, organizations and authorities should be long-term oriented, circular, and non-linear, on top of the regular thinking.

For the decision-making triangle, the "considering?" triangle relates to type I, II, and III events. Thus, when exploring, identifying and mapping information, options and preferences from the viewpoints of academia, organizations and authorities, focus should be on all three types of events. Amongst others, this implies for example that external domino effects should be taken into account, and information, options and preferences should be developed regarding such events, even if they are highly improbable.

The way to deal with the risk triangle for gathering and drafting information, options and preferences on each of the blocks (i.e., possible hazards, possible losses, and possible exposure), should be according to the law of iterated expectations, scenario building, and the uncertainty principle. The law of iterated expectations simply states that if an event (e.g., an accident) can be expected somewhere in the future *with a certain probability*, then the event may also be expected at present. Following this principle implies that measures should be taken for preventing some events, among them some of the type II and type III events, as if they might happen now. Of course, taking prevention measures depends, amongst others, on the available budget and on the levels of uncertainty. Scenario building and the uncertainty principle (following the precautionary principle reasoning) are both explained above.

People, procedures and technology (forming the organizational culture triangle) each need to be considered by an operational, a tactic and a strategic manner of thinking, whereby hazards, losses, exposure, opportunities, and profits should be considered within every domain (people-procedures-technology).

9.4.2 The "results?" layer of the risk governance model

For each of the triangles of the loop, a "results?" triangle can be drafted. The risk policy triangle leads to concrete and abstract ideas, rules and guidelines, and laws. The decision triangle leads to knowledge and information about the past, the present and the future. The risk triangle transpires into risk averse, risk taking and risk neutral items of consideration. The organizational culture triangle leads to risk management know-how, risk management know-what, and risk management know-why in each of its domains. The loop of continuous improvement is closed by more concrete and abstract ideas, rules and guidelines, and laws.

9.4.3 The risk governance model

If the triangles from the policy-decision-risk-culture rad, the considering? triangles and the results? triangles are displayed in an integrated way on the same ▶Fig. 9.4, we create a model that can be used to continuously advance the optimization of risk decisions and risk expertise within and across any organization(s).

> Similar to the fact that people do not want physicians only to be able to recognize well-known and "casual" diseases, and not be able to detect a very rare disease, organizations should not be satisfied with risk experts and risk decisions only tackling well-known and "usual" (mostly occupational) risks, and not considering out-of-the-ordinary-thinking risks (such as type II and type III risks) or not using the proper method and the proper data and expertise to tackle certain risks or certain types of risks. In managing risks, to elaborate sustainable solutions, it is often much more important (and more difficult) to identify and define in detail the problem(s), than the solution(s). This can only be achieved by using a risk governance model such as displayed in ▶Fig. 9.4, integrating all possible viewpoints from diverse stakeholders, risk subjects, methodologies and methods, approaches, disciplines, etc. This model strives to make risk decisions truly more holistic, more systematic, more objective and more justified.

9.5 A risk governance PDCA

Risk governance should really be seen as a product of people. Risk decision results depend more than ever on the efforts made by management, stakeholders and employees. In order to make continuous progress, the commitment of these three clusters of people is indispensable. These lines of thought are the basic ingredients of a risk governance PDCA, which is visualized in ▶Fig. 9.5. In agreement with traditional management systems, the idea of continuous improvement is depicted by the circular outline of the figure.

▶Fig. 9.5 shows five inherent and characterizing components of a genuine risk governance policy connected with each other. The combined knowledge and commitment from central management and relevant stakeholders forms the foundation of a strategic and genuine risk governance policy. This should be evident for company employees. Only after this fulfilment is realized, an organization becomes able to concretize its strategy into operational actions and procedures. An accurate evaluation of these actions and their results should optimally result in more knowledge and a higher level of commitment on the part of management and stakeholders. All this can be achieved

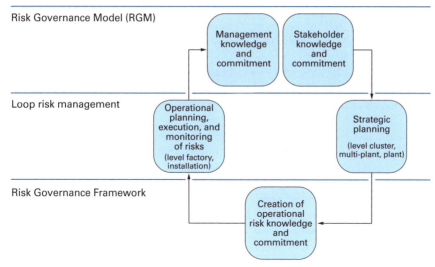

Fig. 9.5: Illustration of a risk governance PDCA.

by using the RGM as explained in the previous section. From this point of view, the upper sphere of knowledge and commitment can be interpreted as a reservoir that is being filled permanently with the essential fuel to undertake a new and more detailed run through the risk management system. Risk governance should in this context be interpreted as a gradual and endless process.

The huge contribution that is dedicated to the human influence in the conceptual and paradigmatic created risk governance cycle is a well-considered choice. Attitudes of people function as influencing factors within the cycle.

The interaction between attitudes of management and stakeholders should result in the choice of a limited number of strategic risk governance-pillars at the cluster, multi-plant and/or plant-level, which need to form the foundations of the risk governance policy. Such pillars lead in turn to a risk governance framework that determines the limits of the risk governance policy and that enables to encompass all the risk management and risk governance efforts by the company.

Once the prospective risk governance policy is written, it has to be used to gain an understanding of the risks that have to be managed. Risks also need to be communicated to all the internal employees, as the understanding of risks together with the commitment with respect to risk governance can be interpreted as crucial factors towards an increased risk decision performance. On the one hand, this process should occur from a top-down approach. Already in the selection process, management can decide to use the risk governance mentality of potential employees as a distinguishing parameter. Subsequently, after an employee's commencement of employment, he/she should still be informed regularly by central management or by risk management about the risk governance course it takes, by means of frequent information sessions. On the other hand, a simultaneous bottom-up approach is required. Employees must feel that they dispose of a key that opens the gate towards an operational fulfilment of the chosen risk governance strategy. The design of some motivation programs or suggestion systems, which enable employees to spread their ideas, e.g., on a central intranet, is a concrete example of such a bottom-up approach.

Once employees are informed and convinced of the importance of a structured risk governance policy, top and middle management together with risk management can proceed to develop and implement an operational planning of specific activities and instructions.

Risk governance can be described as a black box, which even though it is fed with essential input by the management team and by risk managers, basically depends on people's efforts. Every person should therefore evolve towards a self-guided leader.

The application of consistent key performance indicators is a prerequisite, which can be considered as an ever returning issue within a long-term risk governance policy and which enables organizations to make their efforts measurable. An eventual underestimation of the importance of indicators within an organization makes it impossible to examine whether realized actions have led to a significant progress. The operational information and scores on indicators that become available after the completion of the integral conceptual management system can be compared with the essential fuel to drive the internal motor of the risk management system. Realized scores on indicators should be seen as the material that drives an organization forward to new achievements and sharper objectives. In this context, corporate risk governance should really be seen as a gradual process, which can never be ratified as being finished.

The completion of the integral conceptual management system may lead to the formal announcement of obtained results and desired actions for the future, based on information that can be filtered out of the circle. Management and the most prominent stakeholders can choose to publish this kind of information, as a documented proof of significant efforts and realized results.

9.6 Risk governance deficits

IRGC (International Risk Governance Council) defines risk governance deficits as "deficiencies or failures in the identification, assessment, management or communication of risks, which constrain the overall effectiveness of the risk governance process". Understanding how deficits arise, what their consequences can be and how their potential negative impact can be minimized is a useful starting point for dealing with emerging risks as well as for revising approaches to more familiar, persistent risks (15).

> Risk governance deficits operate at various stages of the governance process, from the early warnings of possible risk to the formal stages of assessment, management and communication. Both underestimation and over-estimation can be observed in risk assessment, which may lead to under-reaction or over-reaction in risk management. Even when risks are assessed in an adequate manner, managers may under- or over-react and, in situations of high uncertainty this may become clear only after the fact. Human factors influence risk governance deficits through an individual's values (including appetite for risk), personal interests and beliefs, intellectual capabilities, the prevailing regulations or incentives, but also sometimes through irrational or ill-informed behavior (15).

9.6 Risk governance deficits

IRGC defined two clusters identifying the causes of the most frequently occurring risk governance deficits:

- The assessment and understanding of risks (including early warning systems).
- The management of risks (including issues of conflict resolution).

For the first, they identified ten deficits that can arise when there is a deficiency of either scientific knowledge or knowledge about the values, interests and perceptions of individuals and organizations.

1. The failure to detect early warnings of risk because of missing, ignoring or exaggerating early signals of risk.
2. The lack of adequate factual knowledge for robust risk assessment because of existing gaps in scientific knowledge, or failure to either source existing information or appreciate its associated uncertainty.
3. The omission of knowledge related to stakeholder risk perceptions and concerns.
4. The failure to consult the relevant stakeholders, as their involvement can improve the information input and the legitimacy of the risk assessment process (provided that interests and bias are carefully managed).
5. The failure to properly evaluate a risk as being acceptable or unacceptable to society and the failure to consider variables that influence risk acceptance and risk appetite.
6. The misrepresentation of information about risk, whereby biased, selective or incomplete knowledge is used during, or communicated after, risk assessment, either intentionally or unintentionally.
7. The failure to understand how the components of a complex system interact or how the system behaves as a whole, thus a failure to assess the multiple dimensions of a risk and its potential consequences.
8. The failure to recognize fast or fundamental changes to a system, which can cause new risks to emerge or old ones to change.
9. The inappropriate use of formal models as a way to create and understand knowledge about complex systems (over- and under-reliance on models can be equally problematic).
10. The acknowledgement that understanding and assessing risks is not a neat, controllable process that can be successfully completed by following a checklist. Failure to overcome cognitive barriers to imagining events outside of accepted paradigms ("black swans" or type III events)

For the second cluster (management of risks) they identified 13 deficits related to the role of organizations and people in managing risks, showing the need for adequate risk cultures, structures and processes (15):

1. The failure to respond adequately to early warnings of risk, which could mean either under- or over-reacting to warnings.
2. The failure to design effective risk management strategies that adequately balance alternatives.
3. The failure to consider all reasonable, available options before deciding how to proceed.

4. Inappropriate risk management occurs when benefits and costs are not balanced in an efficient and equitable manner.
5. The failure to implement risk management strategies or policies and to enforce them.
6. The failure to anticipate the consequences, particularly negative side effects, of a risk management decision, and to adequately monitor and react to the outcomes.
7. An inability to reconcile the time-frame of the risk issue (which may have far-off consequences and require a long-term perspective) with decision-making pressures and incentives (which may prioritize visible, short-term results or cost reductions).
8. The failure to adequately balance transparency and confidentiality during the decision-making process, which can have implications for stakeholder trust or for security.
9. The lack of adequate organizational capacity (assets, skills and capabilities) and/or of a suitable culture (one that recognizes the value of risk management) for ensuring managerial effectiveness when dealing with risks.
10. The failure of the multiple departments or organizations responsible for a risk's management to act individually but cohesively, or of one entity to deal with several risks.
11. The failure to deal with the complex nature of commons problems, resulting in inappropriate or inadequate decisions to mitigate commons-related risks (e.g., risks to the atmosphere or oceans).
12. The failure to resolve conflicts where different pathways to resolution may be required in consideration of the nature of the conflict and of different stakeholder interests and values.
13. Insufficient flexibility or capacity to respond adequately to unexpected events because of bad planning, inflexible mindsets and response structures, or an inability to think creatively and innovate when necessary.

Diagnosis and remedy of deficits is not a one-time event, but rather an on-going process of finding problems and fixing them. It relies on an interactive process between risk assessment and management, and between risk generators and those who are affected by risks.

9.7 Conclusions

Risk governance cannot take place in isolation, as argued by Renn (1). Risk governance cannot be applied in a standard way in all locations, political cultures, organizations, and risk situations. It should be open to flexibility and adaptation in order to reflect the specific context of risks. In modern societies, a myriad of risk-influencing factors come into play when considering the wider environment of risk governing. To build the adequate capacity for successful risk governance at all levels, an organization may employ the risk management system, including standards and guidelines (operational), the risk governance framework (tactic), and the risk governance model (strategic), as elaborated and explained in this chapter.

References

1. Renn, O. (2008) Risk Governance. Coping with Uncertainty in a Complex World. London, UK: Earthscan Publishers.
2. Hilb, M. (2006) New Corporate Governance. Berlin, Germany: Springer.
3. Fuller, C.W., Vassie, L.H. (2004) Health and Safety Management. Principles and Best Practice. Essex, UK: Prentice Hall.
4. International Risk Governance Council (2006) White paper on Risk Governance towards an integrative approach. Geneva, Switzerland: IRGC.
5. Reniers, G.L.L. (2010) Multi-plant Safety and Security Management in the Chemical and Process Industries. Weinheim, Germany: Wiley-VCH.
6. Oil and Petrochemical Industry Technical and Safety Committee (2001) Code of Practice on Safety Management Systems for the Chemical Industry. Singapore: Ministry of Manpower.
7. Health and Safety Executive (1993) Successful Health and Safety Management, HS(G)65. Sudbury, UK: HSE Books.
8. Reniers, G. (2012) From risk management towards uncertainty management. In: Risk Assessment and Management. Zhiyong, editor. Cheyenne, WY: Academy Publishing.
9. O'Riordan, T., Cameron, J. (1996) Interpreting the Precautionary Principle. London, UK: Earthscan Publishers.
10. Senge, P. (1992) De Vijfde Discipline, De kunst en de praktijk van de lerende organizatie. Schiedam, The Netherlands: Scriptum Management.
11. Bryan, B., Goodman, M., Schaveling, J. (2006) Systeemdenken, Ontdekken van onze organizatiepatronen. Den Haag, The Netherlands: Sdu Uitgevers.
12. Aven, T. (2010) Misconceptions of Risk. Chichester, UK: John Wiley & Sons.
13. Tversky, A., Kahneman, D. (2004) Loss aversion in riskless choice: a reference-dependent model. In: Preference, Belief, and Similarity: Selected Writings. Shafir, editor. Cambridge, MA: MIT Press.
14. Reniers, G. (2012) Integrating risk and sustainability: a holistic and integrated framework for optimizing the risk decision and expertise rad (ORDER). Disaster Adv. 5:25–32.
15. International Risk Governance Council (2009) Risk Governance Deficits: An Analysis and Illustration of the Most Common Deficits in Risk Governance. Geneva, Switzerland: IRGC.

10 Examples of practical implementation of risk management

This chapter will illustrate, through several examples, how to put engineering risk management into practice. We will focus on "research and teaching" topics as it is more complex and more challenging to implement the "engineering risk management" principles and methods discussed in this book, in such research environments, where a continuous evolution and constant changes of the risks and working environment and of the production processes (e.g., lab syntheses) are taking place. As already indicated in earlier chapters, safety management (as part of risk management) is a quality system used to encompass all aspects of safety throughout an organization. It provides a systematic way to identify hazards and control risks while maintaining assurance that the risk controls are effective. It is a global challenge for each organization to establish a safety management program and plan.

Research and teaching within certain fields of knowledge and science have an array of unique hazards that reflect both the variety and the continuous evolution of their operations. These hazards include chemical, physical, biological and/or technical facets. For example there is an increasing awareness of reactive chemistry hazards. While controlling these hazards is frequently accomplished through engineering approaches such as ventilation and procedures, the long history of repeated incidents suggests that a more formal approach to hazard recognition and management is required.

Academia is composed of many different actors: scientific staff, researchers, teachers, technicians, students, apprentices, administrative staff, short-term visitors, external stakeholders, etc. Those people have different skills, education and knowledge. Hence an overall safety management approach should address the different requirements needed by the diverse population.

Research activities have become more complex over the last decades with more interrelationships and interdependencies. Moreover, new technologies and innovations in developing new materials introduce new risks. This complexity, combined with increasing multifunctional use of space and also increasing population densities with high turnover, creates larger risks to society (and to the research community) while at the same time their acceptance is decreasing. Public perception is generally years behind current practices and reality.

Many risk analysis techniques and risk management emerged in the industry from the 1960s onwards. This can be regarded as a reaction to some major accidents, as well as the desire to achieve higher performance, improve production, quality, and workers' health. Often regarded as centers of conceptualization and theoretical modeling, high schools and universities – the academia/research in a broad sense – are hardly comparable to the industry regarding safety management. The academic world remains also the headquarters of experiment validation associated with a concept of free research. This makes it an environment particularly prone to risk. Indeed, experiments have not always been carried out without incidents or accidents. As observed recently, many more accidents have happened in academia but only very few are reported in the open literature:

- 2006, Mulhouse (France) – Explosion (followed by a fire) in the University's chemistry building. As a consequence, one fatality and several injured (1).
- 2008, Delft (Netherlands) – Fire caused by a short circuit at the Technical University causing considerable financial losses (2).
- 2009, UCLA, Los Angeles (USA) – Explosion (followed by a fire) in the University's chemistry building. As a consequence, one fatality (3).
- 2010, Texas Tech University (USA) – A student received severe burns and lacerations to his face and hands when a mixture of nickel hydrazine perchlorate exploded in a chemistry department laboratory (4).
- 2011, Yale, New Haven (USA) – A student was killed in a chemistry lab by being pulled into a piece of machine-shop equipment (5).

One of the most important factors is that "engineering risk assessment" should be built into scientists' routines. As illustration, each chemical substance used comes with a list of potential risks and appropriate safety precautions through the material safety data sheet (MSDS), although unpredicted toxicity can affect even the most careful chemist. According to Peplow and Marris (1), it seems clear that academic labs are more dangerous than those in industry, because of their more relaxed approach towards safety.

Despite the awareness about the growing risks in the academic/research world, risk management in this environment is even more complex compared to industry because of certain inherent specificities. Moreover, management of change, as expressed by Langerman (6), is even more critical as research laboratories are undergoing continuous and rapid changes. Furthermore, teaching laboratories are occupied by inexperienced operators who are being exposed to new situations. Existing methodologies for risk assessment are hardly directly applicable.

> Implementing a risk management concept in complex systems requires a deep understanding of the possible risk interactions. Complex systems reside not only in a multifaceted combination but also with the difficulty that they evolve in the uncertainty region. This means that we have to deal with scarce or missing information and nonetheless evaluate the risk that we might face.

10.1 The MICE concept

Langerman (7) discussed the "lab process safety management" (PSM) approach for chemical labs, which was designed to help define when changes need to be handled in a coordinated and structured manner. This methodology is mainly process-oriented and does not satisfy the global/overall approach of how global safety management processes should be implemented in a research or teaching dedicated environment. Eguna et al. (8) presented some comments about the management of chemical laboratories in developing countries. An initial safety audit revealed plenty of room for improvement. Therefore an eight-session workshop was conducted for the laboratory personnel over a period of 8 weeks.

These two examples, among others, indicate that there is plenty of room for the implementation of global safety and risk management in research and teaching institutions. One solution is the implementation of a safety management program called MICE (management, information, control and emergency) based on a solid education, adapted to the target audience. This program, based on four levels, is similar to the Deming wheel process or the improved plan-do-check-act as described by Platje and Wadman (9). The four components of the MICE concept are (10,11):

- **M** – The management step
- **I** – The information and education step
- **C** – The control step
- **E** – The emergency step

10.1.1 The management step

The management step concerns different topics such as:

- The welcoming and training of new collaborators (every collaborator, independently from its activity, or student going to practical labs, should have a course introducing them to the basics of safety, fire-fighting training and first aid).
- The decentralized safety management and organization where each research and teaching unit has a safety delegate or coordinator (acting as a first-line safety actor).
- Lab-door panels (including information on present hazards, responsible and contact persons, prohibitions and requirements, safety classification, cleaning issues, etc.) on every research and teaching lab.
- The hazard mapping of all research/teaching labs and offices allowing identifying laboratories with a high-level of danger or cumulative hazards.
- Near-miss, incident and accident web-based interface and database allowing for analyzing and implementing adequate corrective measures in order to avoid the event's repetition.

10.1.2 The information and education step

The information part of the MICE program is mainly related to targeted education or workshops for students (bachelor, master or PhD students), co-workers, researchers, technicians, teachers, administrative and technical staff as well as to external contractors. Web sites especially dedicated to safety should be developed including a comprehensive online safety manual, tutorials on different hazards that collaborators could face in their activities, training videos on how to behave in case of emergency, how to deal with special hazards or how to safely operate in chemical labs, and where someone could find help from a safety specialist. Emergency equipment and their use should also be depicted with training videos, operating manuals and directives.

Newsletters, information panels or paper information could be used as extra communication tools; however we should not forget that nowadays everyone is submerged with papers and emails. A proactive response should be implemented instead of passive communication means.

10.1.3 The control step

Every management process needs a control step. This could be realized by safety audits of each research and teaching lab in order to ensure that the minimal safety requirements are satisfied. This ensures that the management of any change is always covered and mastered. Effectively, as process and procedures are rapidly evolving in research, we have to make sure that adapted safety management is equally reactive and proactive. These audits also have an educational issue as they should be realized in the presence of the individual unit safety coordinator explaining the observed deviation and remediations to be implemented. It allows for rapidly accessing the involved risks and implementing the adequate corrective measures in terms of prevention and protection.

10.1.4 The emergency step

The final step is the emergency step. Despite what one could imagine, emergency is not entirely related with professional intervention squads such as firemen, first-aid, and technicians, but also on the education on how to behave correctly in case of accident (call the center, evacuation drills, behavior, rules, first intervention, training, etc.) and also on how to act after intervention squads have left. In our opinion, emergency is also concerned with remediation and how to recover from a physical or material damage.

> The MICE concept allows for the implementation of a safety management system covering the management step, the information and education, the control part and finally the emergency. It is comparable to a quality management where all the facets have to be covered in order to implement an appropriate process.

10.2 Application to chemistry research and chemical hazards

Research and teaching labs can be too crowded, and such overcrowding raises the risk of spills. Waste disposal becomes a major issue. Most chemistry labs have open bottles where solvents are dumped along with the black gunk left from failed reactions.

Chemical management is a crucial step in order to ensure safety. Obviously, many chemicals should not be manipulated without any confirmation that the chosen equipment is adapted for the purpose and that workers are correctly trained, especially regarding the manipulation of carcinogenic, mutagenic, reprotoxic substances (CMR), highly toxic compounds or substances with high energetic reactivities. In order to validate the safety measures at the workplace, safety and health professionals must have access to the information about the substances used throughout the laboratories. Management

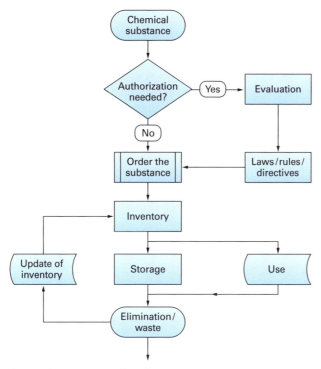

Fig. 10.1: Chemical management flowchart.

of chemicals must find a process that ensures staff have a safe work environment without impairing their innovative thinking. To address this issue, imagine a comprehensive chemical management flowchart starting from the ordering of chemicals and ending with the waste disposal (see ▶Fig. 10.1).

1. Ordering chemicals and substances subject to authorization.
   - Chemical management must start at the ordering stage. All substances must be bought through chemical stores responsible for general negotiations with suppliers, for checking compliance between ordering and shipping as well as for reporting every chemical into the inventory database (see below). A special treatment should be devised for substances that lead to serious health concerns, in particular class 1 CMR's or highly toxic substances. Co-workers must get an authorization to use these chemicals based on a comprehensive work conditions analysis. This quality process allows them to get support from specialists regarding the possibility of replacing all applicable substances with less problematic ones, to verify if safety measures and operating procedures are sufficient, adequate and adapted for the planned project and eventually to determine the need of monitoring measures.
2. Inventory and storage
   - All chemicals have to be inventoried independently of their physical state (solid, liquid or gas) in a dynamic central database (intranet interface),

taking into account chemical information (quantity, purity, MSDS), as well as logistical information (storage place, owner, date). Furthermore, the inventory allows the users to know if a substance is already present and could be borrowed for an initial test. Moreover, the computerized inventory is a powerful tool, enabling safety and health specialists to search for all chemicals with a same specific hazard statement or to prevent aging degradation by checking storage duration.
- Storage must be considered as a first-line safety measure to prevent undesired events. According to CLP regulation [EC implementation of GHS, EC regulation (12)] substances have to be stored in an appropriate manner, taking into account chemical compatibility and segregation depending on their intrinsic reactivity. This part of quality management should also be verified by workplace audit controls.
3. Waste management
- At the end of the process, wastes are treated in a similar manner as pure compounds. They are separated by chemical compatibilities and properties. Self-reactive mixtures have to be rendered inert, as reactive mixtures should not be mixed with incompatibles ones. Once correctly conditioned, they are collected by disposal contractors.

> Chemical management is a major issue due to the fact that a chemical substance does not stop to be active during its whole life. Wastes, representing the end of their life, are still active and thus representing even more hazards compared to pure compounds. In fact, wastes turn into mixtures of unknowns, due to the continuous composition evolution by successive adding. One should not forget that till their destruction (by incineration or physical treatments), chemicals remain active; they are designed for and used because of their activity.

10.3 Application to physics research and physics hazards

There are many hazards related to physics: high magnetic fields, ionizing and non-ionizing rays, cryogenics, lasers, noise, work in hot or cold environment, engineered nano objects, electricity, etc. It would be illusory to draw a comprehensive list of those activities as they are as diverse as imagination could bring. Moreover, often these technologies are combined or linked together, leading to even more complex risk management (13). It this therefore important to have a rapid hazard assessment method that could be incorporated into a risk management process.

Let us illustrate these concepts by applying a risk management process to the use of cryogenics fluids. Cryogenic liquids are liquefied gases that are kept in their liquid state at very low temperatures. They have boiling points below $-150°C$ (e.g., boiling point of helium and nitrogen are, respectively, $-269°C$ and $-196°C$) and are gases at normal temperatures and pressures. These gases must be cooled below room temperature before an increase in pressure can liquefy them. Different cryogens become liquids under different conditions of temperature and pressure, but all have two properties in common: they are extremely cold, and small amounts of liquid can expand into very large volumes of gas (e.g., the ratio volume of gas to volume of liquid at 1 bar and 15°C is 738 for helium and 1417 for neon). However some of them could be flammable (e.g., hydrogen, methane, ethylene and ethane) and/or toxic (e.g., ozone, carbon monoxide and fluorine), adding another dimension to the above-mentioned risks.

The vapors and gases released from cryogenic liquids also remain very cold. They often condense the moisture in air, creating a highly visible fog. In poorly insulated containers, some cryogenic liquids actually condense the surrounding air, forming a liquid air mixture. With the exception of liquid oxygen, which is light blue, all the liquid cryogens are colorless. The properties of many cryogens of being colorless and odourless make them impossible to detect and discriminate by eye or by the sense of smell.

Everyone who works with cryogenic liquids (also known as cryogens) must be aware of their hazards and should know how to work safely with them. Before applying any strategies of risk management we has to understand which hazards we might face.

10.3.1 Hazards of liquid cryogens

Cryogens' hazards can be summarized in two categories: health hazards and hazards related to material properties, refrigerant properties and condensation mechanisms.

The health hazards could be listed as:

- Cold burns and frostbite – Exposure of the skin to a cryogenic liquid or its cold vapour/gas can produce skin burns similar to heat burns. Frostbites are caused by prolonged exposure of unprotected skin to cold vapors or gases. Once tissues are frozen no pain is felt and the skin appears waxy and of yellowish color.
- Contact with cold surfaces – If unprotected skin comes into contact with cold surfaces, like uninsulated pipes or vessels, the skin may stick and flesh may be torn off on removal.
- Effect of cold on lungs – Patients suffering from bronchial asthma or chronic obstructive lung diseases often experience aggravation of bronchospasm on exposure to cold environment. Inhalation of cold mist, gases or vapors from the evaporation of cryogenic liquids worsens the degree of airway obstruction in sensitive patients. Short exposure creates discomfort even in normal subjects and could damage the lungs in case of prolonged exposure.
- Hypothermia – Exposure to low air temperatures can cause hypothermia. Hypothermia is a condition associated with the decrease of body temperature below 35°C. The susceptibility of a person to hypothermia depends upon the temperature, the exposure time and the individual concerned (older people are more likely to succumb). When the body temperature goes below 33°C the victim could fall unconscious or asleep, and after some time could fall into coma.
- Asphyxiation – As mentioned above, cryogenic liquids have a very large expansion rate. The liquid evaporates in such a large volume of gas that displacing air can considerably reduce the amount of oxygen available. Neon, hydrogen and argon have the highest expansion rates. Argon and nitrogen are heavier than air and tend

to accumulate in low-lying areas such as pits and trenches. They can also seep through porous materials, fissures in the soil, cracks in concrete, drains and ducts. The cold vapors may collect and persist in confined spaces creating an oxygen deficiency hazard.
- Toxicity – The acute or chronic exposure of the considered substance will induce health effect depending on the concentration and the exposure duration. This effect is not specific to cryogens and more related to chemical specificities.
- Thermal radiation burns – Exposure to thermal radiation caused by the combustion of flammable cryogenic gases can produce first-, second- or third-degree burns. The degree of severity of a burn depends on the combustion temperature of the gas-air mixture, the distance of the victim from the heat source and the time of exposure.
- Blast/explosion injuries – A blast is a consequence of a confined combustion of a flammable gas. A leak of a flammable cryogenic fluid in a confined space is extraneously dangerous because the flammable gas-air mixture would tend to cumulate in the area. If the gas mixture is ignited, a large overpressure is produced with dramatic consequences.

Hazards related to material properties, refrigerant properties and condensation mechanisms could be listed as:

- Brittle fractures – Some properties of materials can change significantly with temperature. Special care should be taken in selecting the material for equipment employed at cryogenic temperatures because of the risk of brittle fractures. A ductile material if placed in tension will stretch. For low values of tension, the elongation is proportional to the tension, and as the stress is removed the material returns to its original length (elastic behavior). For higher values of tension, the elongation is permanent. Eventually the material will break at a maximum tension characteristic of the material, called "tensile stress" or "ultimate stress". A brittle material does not exhibit this behavior, it rather suddenly breaks at high enough tension with no permanent deformation observed prior to rupture. As temperature is lowered some materials undergo a change from ductile to brittle behavior. Fractures and cracks cause spills and leakages of the cryogenic liquid and gas.
- Thermal contraction leakage – Most materials have a positive thermal expansion coefficient meaning that when they warm up they expand and when cooled down they contract. Pipes, vessels and joints employed with cryogens must be of materials carefully selected according to their thermal expansion coefficient. Cooling down a material from room temperature to cryogenic temperatures causes a significant thermal contraction. For example, when going from ambient to cryogenic temperatures iron-based alloys contract about 0.3% while many plastics contract well over 1%. A very large stress is produced at the joints of rods or pipes of a cryogenic system if these are not allowed to contract freely when cooled down. Such stresses might result in broken joints or cracks along the pipe that would produce gas or liquid leakages.
- Overpressure – As mentioned above, cryogenic liquids vaporize into large volumes of gas. The normal heat inlet through the insulated walls and pipes of the storage vessel raises the temperature of the cryogenic liquid. For a steady heat flow the

liquid boil-off rate of the vessel could be determined by the amount of liquid that slowly vaporizes as temperature rises. As the liquid vaporizes the pressure inside the vessel increases. In this situation overpressure could be caused by failure of the protection systems such as relief valves or rupture disks or the loss of thermal insulation. If the thermal insulation is damaged, for example if the vacuum in the vacuum jacket of the vessel is compromised, this could produce a large boil-off rate that the relief valves could not manage. Pressure can rise extremely high when cold liquid and vapor are stored in pressurized vessels that are not adequately vented and when refrigeration is not maintained. A pressure build-up may produce a burst with the sudden release of large quantities of cold liquid and gas as well as projection of mechanical parts.

- Combustion caused flammable cryogens – Deflagration occurs when a portion of a combustible gas-air mixture is heated to its ignition temperature. As the combustion starts the heat released is sufficient to ignite the adjacent gas, producing the propagation of the flame front through the combustible mixture. For example the deflagration velocity of hydrogen is 2.7 m/s.
- Combustion caused by oxygen-enriched atmosphere – Special care should be taken when using liquid oxygen as coolant. Oxygen is not itself flammable but supports combustion. Oxygen is heavier than air and tends to accumulate in low-lying areas such as pits and trenches, and can also seep through fissures in the soil, cracks in concrete, drains and ducts. The cold vapors may collect and persist in these confined spaces, creating an oxygen-enriched atmosphere. Personnel should not enter areas where the atmosphere is rich in oxygen (>22%). Hair, clothing and porous substances may become saturated in oxygen and burn violently if ignited. Clothes contaminated with oxygen should be kept for at least 15 minutes in the open air. Combustible material in presence of oxygen-enriched atmosphere ignites more easily, burns much more vigorously and can react explosively. Moreover, materials normally non-combustible in air such as stainless steels, mild steel, cast iron and cast steel, aluminium and zinc become combustible in oxygen-enriched atmosphere.
- Condensation hazards – Because of their very low temperature, almost any cryogen can solidify water or carbon dioxide. The presence of solid particles within a fluid system can cause damage to the system and hazardous situations such as overpressure and leakages. The solid particles can erode valves and gaskets producing leakages. If a large amount of solid particles cumulate inside pipes or in proximity to relief valves, these could be blocked and prevent gas to be released. If gas is trapped inside the system, pressure could increase to dangerous levels. Another issue related to the condensation of air is the phenomena of oxygen-enrichment. The development of an oxygen-enriched atmosphere is a fire and explosion hazard. Air is composed of 21% oxygen and 78% nitrogen. Liquid air produced by condensation on cold surfaces does not have the same composition of the vapor being condensed. Because of the higher boiling point of oxygen (90.2 K) than nitrogen (77.4 K), oxygen condensates preferentially to nitrogen. It will produce liquid air with an oxygen concentration that could reach 50%. Liquid air is therefore very rich in oxygen and an explosive hazard is more present.

10.3.2 Asphyxiation

We do not intend to discuss all the hazards linked with the use of cryogens, for such a discussion would fall out of the scope of this book. We will concentrate on asphyxia, caused by the oxygen content lowering, and discover how to apply the risk management process to this particular situation.

Ventilation is a key issue when handling and storing cryogenic liquids. Large quantities of cryogenic liquids must be stored in open air or in well-ventilated areas. Small, not-ventilated rooms should be avoided to prevent build-up of the gas as the cryogen evaporates. A well-ventilated laboratory has a number of air renewals per hour between five and ten. But what will happen if some litres of a cryogen leaks into the room? Is there a health problem? What could be the consequences?

In order to answer those questions, trials were held on a real scale where several litres of liquid nitrogen were poured on the floor and the evolution of the oxygen content was measured at different locations and heights in the room. Results indicate that they were huge differences depending on the ventilation efficiency. This led to the conclusion that an oxygen detection system should be installed if the quantity of stored cryogen in the room is above 0.4 l/m^3 of space when correctly ventilated and above 0.3 l/m^3 for non-ventilated rooms. These values correspond to the observed limits where the oxygen content was below 19% for more than 3 minutes at a height of 1.60 meters (being the height-average of the human face). An oxygen detector should be installed depending on the quantity of liquid cryogen stored in the room. The detector is equipped with a visual alarm that becomes active if the oxygen level goes below 19%. If the oxygen level drops below 18% an acoustic alarm is also activated indicating the evacuation of the room. The instrument should undergo regular calibration checks and routine maintenance to ensure reliable performance. Moreover, a clear labeling of the room door should advise people entering that this room may have potential asphyxiation problems when the alarm is on (see ▶Fig. 10.2).

This example exhibits that even in the absence of regulation it is possible to implement a risk management strategy by replacing rules by experiments and interpreting them to define internal regulation that could be applied in similar situations.

> Most of the physical hazards have in common that they are not correctly evaluated by human senses. Technical measures should inform our senses and mind that we are facing a hazard and that a risk might be present. Often, their actions on human bodies are not directly connected to their use; i.e. cryogens are used to attain low temperatures but one of their drawbacks is the possibility of death by asphyxiation in case of a large release in the environment.

Fig. 10.2: Alarm warning and door panel indication.

10.4 Application to emerging technologies

Emerging technologies have in common that they are evolving in higher levels of uncertainty and complexity, accelerating speed and competency-destroying change. These are among the characteristics that make managing emerging technology distinct from managing established technology. We could then define emerging risks as new or already-known risks that are difficult to assess ("known unknown risks", see Chapter 2). The IRGC goes a bit further by defining three categories of emerging risks (14):

1. Uncertain impacts: Uncertainty resulting from advancing science and technological innovation.
2. Systemic impacts: Technological systems with multiple interactions and systemic dependencies.
3. Unexpected impacts: Established technologies in evolving environments or contexts.

The OSHA (15) definition of emerging risk stipulates that any risk that is new and/or increasing is emerging. By "new" they define:

- The risk did not previously exist and is caused by a new process, new technologies, new types of workplace, or social or organizational change; or
- A long-standing issue is newly considered as a risk because of a change in social or public perception; or
- New scientific knowledge allows a long-standing issue to be identified as a risk.

The risk is "increasing" if:

- The number of hazards leading to the risk is growing; or
- The likelihood of exposure to the hazard leading to the risk is increasing (exposure level and/or the extent of human values exposed); or
- The effect of the hazard is getting worse (severity and consequences and/or the extent of human values affected).

All these definitions have in common that often their acute and chronic impacts are not well known. This constant transformation requires a dynamic approach. We need to scan, monitor and respond to technologies and strategies that are constantly in motion. The main challenge for the future resides in dealing with multiple interacting risks. Most of them will be not sufficiently known or even unknown, leading to remain in the uncertainty zone.

The main question to answer is: "How do we protect against something where we have insufficient information about its consequences?" Knowing what to implement and when to make the change can make anyone wander around in a haze of confusion, but with a few process steps, the management could become much simpler. However, given the risky and often unproven nature of these technologies, considerable confusion exists on how to manage their implementation (16). The key issue is the ability to adequately learn about new advances, boiling down to a 3-step phase.

1. Keep your eyes open for new technologies that might assist you. Read articles and scour technology sites where emerging technologies often are introduced. (see "Mindset", ▶Fig. 2.1, Chapter 2)

2. Evaluate the technology against your strategy. Will this new technology increase efficiency, allow entering new markets, fastening your development or reduce costs? (See "Stakeholders and expertise", Chapter 2.)
3. Ask lots of questions and make yourself the knowledge expert if you think the technology might be useful. Reach out to those who wrote articles on the subjects. Attend specialized conferences in order to make your network broader to learn faster. (cf. "Knowledge and know-how", Chapter 2.)

The second issue is the implementation of these new advances:

1. Outline the benefits and risks of the change. Prepare a cost-benefit analysis that lists what you hope to gain with the change and any risks that might be associated with that change. Sometimes just looking at this list can help make the decision to move forward or drop the project.
2. We often hear that small is beautiful, so start small when possible. Begin with a limited test of the technology. A mini-implementation can help you evaluate new technologies within your own products, processes and services much better than any literature.
3. Establish and document a communication plan. This should include project communications, such as status, timetables, phases, issue resolution and cost. It also should include how you will communicate with employees, or externals assisting with the implementation.
4. Plan a fall-back path. Newer technologies may have "bugs" or would not work as promised. When implementing, be on guard for the unknown. This way, if something comes up that makes the implementation a bad idea, you have a way to scale back to what you had before.
5. With unknowns we must be proactive. It is better to act instead of react, especially when we are driven by the need to find solutions to short-term challenges. Given the current economic climate, the quality of foresight and planning based on longer-term planning are not really en vogue.

Bhattacherjee (17) discussed key organizational factors affecting the implementation of emerging technologies. He concluded that adequate efforts and resources must be devoted to understanding and managing these challenges. IRGC developed four risk governance dimensions. including 11 themes for improving the management of emerging risks. The grouping begins with risk governance (a concern of strategy, top management and organizational design), then moves to an organizations risk culture, to training and capacity building, and finally, to adaptive planning and management (18):

I. Risk governance: strategy, management and organizational matters.
   1. Set emerging risk management strategy as part of the overall strategy and organizational decision-making.
   2. Clarify roles and responsibilities.
II. Risk culture.
   1. Set explicit surveillance incentives and rewards.
   2. Remove perverse incentives to not engage in surveillance.
   3. Encourage contrarian views.

III. Training and capacity building
   1. Build capacity for surveillance and foresight activities.
   2. Build capacity communicating about emerging issues and dialoguing with key stakeholders.
   3. Build capacity for working with others to improve the understanding of, and response to, emerging risks.
IV. Adaptive planning and management.
   1. Anticipate and prepare for adverse outcomes.
   2. Evaluate and prioritize options; be prepared to revise decisions.
   3. Develop strategies for robustness and resilience.

Improving the management of emerging risk requires improvement in the communication to identify and characterize such risks. The transparency is a precondition for having the research and innovation perceived as balanced, fair and beneficial for the society. What is beneficial for the society is, however, not a term precisely defined and it can be a topic of major differences in opinion, depending on the stakeholder groups in society. In particular, the question, "Is a particular innovation or new technology beneficial for the society?" (e.g., engineered nanotechnologies) often cannot be answered in a straightforward way. This mainly implies that the question should be posed in the reversed way: "How can we be sure that the innovation will not involve risks that we do not want to accept?" This is the question that leads to the precautionary principle (better safe than sorry) (18,19). It states that "in the absence of suitable hazard data, a precautionary approach may need to be adopted". But the practical implementation of the general principles poses a lot of challenges and leads to different solutions. When dealing with the uncertainty zone, we do not have strict answers or even questions. Any discussion about emerging risk may start with the question: "While it is emerging, it is not yet a risk, when it is a risk, it is no longer emerging". The concern about emerging risk is magnified by the fact that our knowledge about the phenomenon is incomplete and we are not sure what exactly we are taking about. But should this indicate that nothing has to be done? The answer is clearly "no". The route map when dealing with emerging risks (ER) could be listed as (20):

- Earlier recognition of ER.
- More systematic recognition of ER by evaluating precursors (on web, papers, conferences, debates, etc.) and monitoring their development; identifying similarities with known risk and their precursors (find analogies).
- Better identification of critical ER.
- Recognition of interdependencies and relations among all risks.
- Improving the knowledge in triggers, drivers and factors of ER.
- Set up a monitoring process and follow-up.
- Systematic interlinking among hazards, vulnerabilities and stakeholders.

Emerging technologies have in common that they are evolving in higher levels of uncertainty and complexity, accelerating speed and competency-destroying change. Emerging risks could be defined as new or already-known risks that are difficult to assess as: uncertain impact, systemic impacts or unexpected impacts. Mastering those risks needs some prerequisites such as: keeping the eyes open for new technologies, evaluating them, and asking as many questions as needed to gain knowledge.

10.4.1 Nanotechnologies as illustrative example

As this field is rapidly evolving according to the progress being made worldwide, this section has to be taken as a snapshot of what is the current state-of-the-art.

How do you protect people from materials with properties that are unknown? The approach is similar to training firemen to fight fires from unknown material.

The properties of manufactured nanomaterials (materials made of nanoparticles smaller than 100 nm produced intentionally by humans, ENP) are paving the way for a wide variety of promising technological developments. However, due to the many uncertainties, conventional assessment of the associated hazards and exposure based on quantitative, measurable criteria is difficult. These uncertainties will only be removed as scientific understanding of the properties of nanomaterials advances.

The handling of nanomaterials is a challenge because of the unknowns involved. If there is a known impact, does it arise from only one part of the material distribution? Doing nothing is not acceptable; this indicates that education guidance and handling procedures must be developed. Key elements must be largely disseminated.

Furthermore, given the current state of knowledge on manufactured nanomaterials, it is highly likely that many years will pass before we know precisely which types of nanomaterials and associated doses represent a real danger to humans and their environment. Indeed, the assessment of potential health effects following exposure to a chemical must take into account the extent and duration of exposure, the biopersistence, and interindividual variability, all subjects on which we have practically no knowledge for the field of nanomaterials (21).

It is therefore extremely difficult to conduct a quantitative risk assessment in most work situations involving nanomaterials with the currently available methods and techniques. It will be challenging – at best.

With unknowns, we must be proactive. In many cases, we do not know what we are looking for! So the question is: "How do we proceed?" While we do not have all the answers, we can take a number of precautions. The focus of our actions must be to:

- Keep ourselves safe.
- Keep our colleagues safe.
- Keep the general population safe.
- Keep the facilities safe.
- Keep the environment safe.

Nano safety is a growing concern that many institutions are working on because nowadays, today's environment requires that people and organizations are responsible for their actions.

We will not discuss all the available information from the literature but we will focus on the management aspects of several different safety methodologies or procedures. We have chosen to compare results using three different methodologies: the tree decision (22), the control banding NanoTool 2.0 (23) and the ANSES method (21). For supplementary methods, we refer to (24–27).

The decision tree developed by Groso et al. (22), concentrates on the physical aspect of the ENP: in powder, suspension or in matrix. The process starts using a schematic decision tree that allows classifying the nano laboratory into three hazard classes similar to a control banding approach (from Nano3 = highest hazard to Nano1 = lowest hazard). Classifying laboratories into risk classes would require considering actual or potential exposure to the nanomaterial as well as statistical data on health effects of exposure. Because these data (as well as exposure limits for each individual material) are not available, risk classes could not be determined. For each hazard level we then provide a list of required risk mitigation measures [technical (T), organizational (or procedural, P) and personal (P)]. This tool is intended for researchers to self-quantify their hazard and risk level. Depending on the nano-hazard classification, several measures are defined. It has to be noted that it is the sole method of defining and proposing measures at the strategic, technical, organizational, and personal levels. Moreover, they also defined adequate measures for visitors, technical and maintenance staff, intervention squads, pregnant woman protection and medical survey.

NanoTool (23) was developed to support first-line occupational health professionals and researchers to evaluate the potential risks related to production and down-stream use of nanomaterials in a research work environment. In the NanoTool, carcinogenicity, mutagenicity and/or reproduction toxicity (CMR), and general toxicity and dermal toxicity of parent material are used as distinct parameters with assigned severity points. In the CB NanoTool, parameters related to emission potential are dustiness/mistiness (substance emission potential) and amount of ENP handled (activity emission potential). The CB Nanotool links the hazard and exposure bands, which have the same ranges of scores, into four risk levels, and consequently to control bands linked to the risk levels.

ANSES (21) is intended to be used by those adequately qualified in chemical risk prevention. It uses the classification of either the bulk material or an analogous substance as the starting point for the hazard banding process if the ENP is not a biopersistent fibre. Hazard parameters such as dissolution time and reactivity may increase the hazard band. ANSES covers emission potential by initial banding based on the physical state of the material, ranging from solid (exposure band 1) to aerosol (exposure band 4). Further modification of the bands (increment) is possible either due to the substance emission potential or the process operations (activity emission potential). In ANSES, the five hazard and four exposure (emission potential) bands are directly linked into five control bands. The hazard band is dominating the allocation of the control band (or risk level) because the highest hazard band, e.g., in case of persistent fibres or lack of information, requires the highest control band independent of the exposure band.

Let us discover, using a very simple example, the outcomes using the three different methods. The illustrative example is the preparation of a wafer on which will be deposited a black ink containing carbon black ENPs. The process could be described as:

1. Carbon black particles (FW200), d =13 nm and specific surface area of 550 m^2/g are received in 500 g containers.
2. Weighting of C black in order to distribute the powder into smaller containers.
3. 30 grams of carbon black is weighted from a small container.
4. Cleaning surfaces surrounding the container and outer walls of container.
5. Add weighted C black to the previously prepared liquid resin.
6. The prepared mixture is stirred in a closed flask to obtain the desired ink.
7. The ink is deposited by pipette on the wafer.
8. The wafer is baked in a hermetically closed oven.

Tab. 10.1: Results of the evaluation using three different evaluation methods.

	Tree method	CB NanoTool	ANSES
Hazard level	Level 3 on the scale of 3	Level 3 on the scale of 4	Level 5 on the scale of 5
Control measures	Technical, organizational and personal measures for Nano 3	Containment of the process	Full containment and review by a specialist

We will concentrate on step 3 as it is the most hazardous in relation to exposure. Using the three methods we end up with the results expressed in ▶Tab. 10.1.

We could note that the classification is rather similar for the three methods, being in the highest hazard zone. However, they largely differentiate when looking at the measures that should be followed. The tree method (22) is largely the most comprehensive, indicating in detail what should be applied, then the NanoTool indicates that the process should be confined, ANSES add to this that a review should be made by a specialist.

We may then raise the following questions:

- Who is the specialist when a lot of unknowns are predominant and when we act in the uncertainty zone?
- How should we apply the precautionary principle?
- How should we implement safety and risk management in such conditions?

Answering those questions is applying risk management principles expressed in the preceding chapters, taking into account the uncertainty zone. It is hard to find one "expert" who is capable on his/her own of addressing complex systems; it should be a multidisciplinary team effort.

When using any kind of methodologies, we should not forget that the ultimate goal is to safely operate a hazardous process. The more information we have, the better will be the outcome.

10.5 Conclusions

If there are so many good models on risk management (see Chapter 3), then why is there still such a high failure rate of projects? It is very easy to acknowledge that there are large risks present when undertaking development projects. It can therefore be tempting for development managers to completely ignore engineering risk management approaches; risk management may imply that the failure of the project is almost a certainty. The key aspects are:

- Focus on sustainability: Leverage existing practices for risk management purposes.
- Be pragmatic: Customized strategy should be supported by simple and efficient methods that meet the needs of, and add value to, managers.
- Take a balanced approach: Balance between level of investment, value expected from investment, and the capacity of the organization.
- Be realistic: The sophistication of the risk management regime must be in step with the maturity of the organization's other management processes.

- Provide leadership: Establish champions across the organization, with clear accountabilities

When you have effective risk management in place, you can focus your planning on avoiding future problems rather than solving current ones. You can routinely apply lessons learned, in order to avoid crises in the future rather than fixing blame. You can evaluate activities in work plans for their effect on overall project risk, as well as on schedule and cost. You can structure important meeting agendas to discuss risks and their effects before discussing the specifics of technical approach and current status.

Above all else, you can achieve a free flow of information at and between all program levels, coordinated by a centralized system to capture the risks identified and the information about how they are analyzed, planned, tracked, and controlled. You can achieve this when risk is no longer treated as a four-letter word, but rather is used in your organization as a rallying perspective to arouse creative efforts.

With effective risk management, people recognize and deal with potential problems daily, before they occur, and produce the finest product they can within budget and schedule constraints. People, work groups, and projects throughout the program understand that they are building just one end product and have a shared vision of a successful outcome.

Engineering risk management should:

- Create value.
- Be an integral part of organizational processes.
- Be part of decision-making.
- Explicitly address uncertainty.
- Be systematic and structured.
- Be based on the best available information.
- Be tailored.
- Take into account human factors.
- Be transparent and inclusive.
- Be dynamic, iterative and responsive to change.
- Be capable of continual improvement and enhancement, etc.

Effective risk management requires the same steps as the decisions encountered at a stop light. Objectives (getting to our destination) must be clear, and attributes of achievement (we must get there before a certain time) must be included. We assess the compliance of others (did everyone stop for the red?) and manage uncertainty based on our risk tolerance (is there time for us to cross on the yellow light?).

Our action/inaction is guided by our analysis and our risk tolerance. Common carriers (railway, bus and air) are very aware that their customers have delegated risk management to them and generally operate very conservatively (by regulation and by choice) so as to ensure that they are not more risk-tolerant than their most risk intolerant client.

References

1. Peplow, M., Marris, E. (2006) How dangerous is chemistry. Nature 441:560–561.
2. Meacham, B., Park, H., Engelhardt, M., Kirk, A., Kodur, V., van Straalen, I., et al. (2010) Fire and Collapse, Faculty of Architecture Building, Delft University of Technology: Data Collection

and Preliminary Analysis. Proceedings, 8th International Conference on Performance-Based Codes and Fire Safety Design Methods, Lund University, Sweden.
3. Kemsley, J. (2009) Learning from UCLA. Chem. Eng. News 87:29–34.
4. Johnson, J. (2010) School labs go under microscope. Chem. Eng. News 88:25–26.
5. Foderaro, L.W. (2011) Yale student killed as hair gets caught in lathe. New York Times April 13th.
6. Langerman, N. (2008) Management of change for laboratories and pilot plants. Org. Proc.Res. Dev. 12:1305–1306.
7. Langerman, N. (2009) Lab-scale process safety management. Chem. Health Saf. 6:22–28.
8. Eguna, M.T., Suico, M.L.S., Lim, P.J. (2011) Learning to be safe: Chemical laboratory management in a developing country. Chem. Health Saf. 6:5–7.
9. Platje, A., Wadman, S. (1998) From Plan-Do-Check-Action to PIDCAM: the further evolution of the Deming-wheel. Int. J. Proj. Man.16:201–208.
10. Meyer, Th. (2012) How about safety and risk management in research and education? Procedia Engin. 42:934–945.
11. Marendaz, J.L., Friedrich, K., Meyer, Th. (2011) Safety management and risk assessment in chemical laboratories. Chimia 65:734–737.
12. European Council (2008) Regulation on classification, labeling and packaging of substances and mixtures. Commission Regulation, (EC) No 1271/2008, Official Journal L 338:0050-0052.
13. Marendaz, J.L., Suard, J.C., Meyer, Th. (2013) A systematic tool for Assessment and Classification of Hazards in Laboratories (ACHiL). Safety Sci. 53:168–176.
14. International Risk Governance Council (2011). Improving the management of emerging risks. Lausanne, Switzerland.
15. European Agency for Safety and Health at Work (2010) European Risk Observatory Report – European Survey of Enterprises on New and Emerging Risks. Managing Safety and Health at Work. Brussels, Belgium: Publication office of the European Union.
16. Day, G.S., Schoemaker, P.J.H., Gunther, R.E. (2000) Wharton on Managing Emerging Technologies, 1st edn. New York: John Wiley and Sons.
17. Bhattacherjee, A. (1998) Management of emerging technologies: Experiences and lessons learned at US West. Inform. Manag. 33:263–272.
18. European Union (1992) Treaty of Maastricht on European Union. Official Journal C 191.
19. Commission of the European Communities (2000) Communication on the precautionary principle, COM 1.
20. iNTeg-Risk FP7 (European Framework Program) project (2012) 4th iNTeg-Risk Conference, "Managing Early Warnings – what and how to look for?" Stuttgart. http://www.integrisk.eu-vri.eu.
21. ANSES (2010) Development of a specific Control Banding Tool for Nanomaterials. Request No.2008-SA-0407.
22. Groso, A., Petri-Fink, A., Magrez, A., Riediker, M., Meyer, Th. (2010) Management of nanomaterials safety in research environment. Part. Fibre Toxicol. 7:40.
23. Zalk, D.M., Paik, S.Y., Swuste, P. (2009) Evaluating the control banding nanotool: a qualitative risk assessment method for controlling nanoparticle exposures. J. Nanopart. Res. 11:1685–704.
24. Van Duuren-Stuurman, B., Vink, S.R., Verbist, K.J., Heussen, H.G., Brouwer, D.H., Kroese, D.E., et al. (2012) Stoffenmanager Nano version 1.0: a web-based tool for risk prioritization of airborne manufactured nano objects. Ann. Occup. Hyg. 56:525–41.
25. California Nanosafety Consortium of Higher Education (2012) Nano toolkit. In: Working Safely with Engineered Nanomaterials in Academic Research Settings. de la Rosa Ducut, J., editor. Riverside, CA: University of California.
26. Schulte, P., Geraci, C., Zumwalde, R., Hoover, M., Kuempel, E. (2008) Occupational risk management of engineered nanoparticles. J. Occup. Environ. Hyg. 5:239–249.
27. Federal Office for Public Health. Precautionary matrix for synthetic nanomaterials. (accessed in 2012) http://www.bag.admin.ch/nanotechnologie

11 Major industrial accidents and learning from accidents

11.1 Link between major accidents and legislation

The tendency of major accidents to force politicians to take political actions ad hoc is illustrated by ►Fig. 11.1, presenting a chronological overview of a non-exhaustive number of significant major accidents worldwide and the chronological developments in a non-exhaustive number of European and US safety regulations.

All political attention comes to focus on the accident that has just happened, while a needed proactive broad political view of the accident prevention issue as a whole is overlooked. The reason for this reactive political behavior is that politicians and company policy makers have limited imaginative power to fully understand the probabilities of accident estimates. Moreover, people are risk averse for gains, but risk taking for losses (see also Chapter 8). This phenomenon causing ad hoc prevention legislation is one of a number of biases in information processing that occurs when we make choices.

Originally described by two psychologists, Daniel Kahneman and Amos Tversky (1), the work has had a considerable impact on economic models of choice. Tversky and Kahneman discovered there is a strong tendency for individuals to be what they called risk averse for gains, but risk taking for losses. The choice experiment has been carried out many times, in many different situations, and the results are very robust. Whether the decision is presented as a loss or a gain will influence what decision is taken. People prefer to hang on their gains, but gamble with their losses. In other words losses and gains are not equally balanced in decision-making. Certain gains are weighted far more heavily than probable losses. As Sutherland et al. (2) point out, this can be easily seen in the balancing of "production" and "safety". Ignoring some safety aspects will lead to almost certain increases in production in the short term. Applying safety measures for highly improbable accidents will lead to additional costs in the short term. Hence, it takes a much larger, and more probable, loss to tip the prevention management decisions in favor of safety with respect to low probability, high-consequence risks.

Nonetheless, the potential for major industrial accidents, which has become more significant with the increasing production, storage and use of hazardous substances, has emphasized the need for a clearly defined and systematic approach to the control of such substances in order to protect workers, the public and the environment.

Major hazard installations possess the potential, by virtue of the nature and quantity of hazardous substances present, to cause a major accident in one of the following general categories:

- The release of toxic substances in tonnage quantities that are lethal or harmful even at considerable distances from the point of release.
- The release of extremely toxic substances in kilogram quantities that are lethal or harmful even at considerable distances from the point of release.
- The release of flammable liquids or gases in tonnage quantities that may either burn to produce high levels of thermal radiation or form an explosive vapor cloud;
- The explosion of unstable or reactive materials.

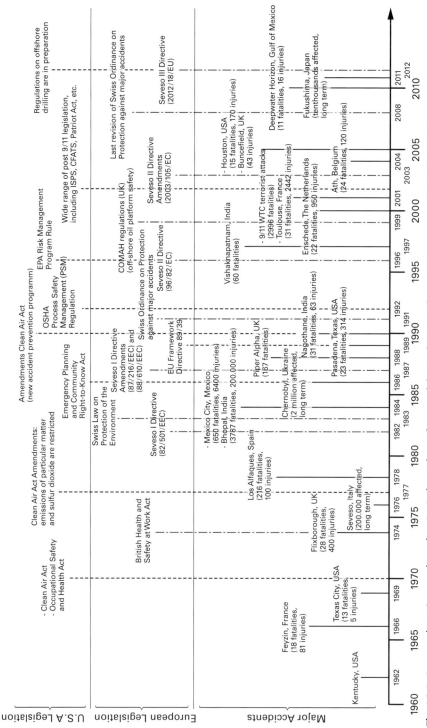

Fig. 11.1: A non-exhaustive number of major industrial accidents and their relation with some European and US safety legislation.

Apart from routine safety and health provisions, special attention should be paid by competent authorities to major hazard installations by establishing a major hazard control system. This should be implemented at a speed and to an extent dependent on the national financial and technical resources available. The works management of each major hazard installation should strive to eliminate all major accidents by developing and implementing an integrated plan of safety management. Works management should develop and practice plans to mitigate the consequences of accidents that could occur. For a major hazard control system to be effective, there should be full cooperation and consultation, based on all relevant information, among competent authorities, works management, and workers and their representatives.

11.2 Major industrial accidents: Examples

Literature sources for the accidents include Wells (3), Reniers (4), and Paltrinieri (5).

11.2.1 Feyzin, France, January 1966

A tank farm including eight spheres containing propane and butane was undergoing a routine drainage operation. The wrong procedure was applied to a 1200 m^3 sphere. The isolating valves became inoperable and an uncontrolled leak from a propane sphere was ignited by a car on a nearby road, flashing back to burn like a flare directly under the sphere. Propane snow had also accumulated within the bund. The refinery fire brigade attempted to put out the fire but ran out of dry foam. The municipal fire company continued to fight the fire using fire water. After 30 minutes, the safety valve lifted and 1 hour later a boiling liquid expanding vapor explosion (BLEVE) ensued. The fire brigade had concentrated on cooling the other spheres and not the burning sphere on the assumption that the relief valve would provide protection. Approximately 340 m^3 of liquid propane was released and partially vaporized, producing a large fireball and an ascending mushroom cloud. The BLEVE killed and injured approximately 100 people in its vicinity.

One debris missile broke the legs of an adjacent sphere, which contained 857 m^3 of propane. A second piece tipped over another sphere containing 1030 m^3 of butane. Another section travelled 240 m to the south and severed all the product piping connecting the refinery area to the storage area. One fragment broke piping near four floating roof tanks. Fires were initiated in this area. Extensive structural damage was caused in the village of Feyzin, about 500 m away. Some 2000 people were evacuated from the surrounding area. A further BLEVE and other explosions occurred as the fire spread. Fire-fighting continued for a further 48 hours until the three spheres that were still intact (and full of propane and butane) were cooled to an appropriate level.

11.2.2 Flixborough, UK, June 1974

In the month of March, a caprolactam plant using six reactors in series, was shut down and reactor "No. 5" was taken out of service. A 20-inch pipe was used to fabricate a dog-leg bypass pipe. The plant was started up on the first of April. On Wednesday 29 May a leak was discovered in the bottom isolation valve on a sight glass fitted to a reactor. The plant was depressurized and cooled down, the leak repaired and the plant restarted. Normal operating conditions of 8.8 bar and 155°C were achieved on 1 June with the plant on hold pending the arrival of high pressure nitrogen needed for the commencement of oxidation. Shortly before 17.00 h, the bypass in place on the No. 5 reactor became unstable with the result that the two bellows units attached to the No. 4 and No. 6 reactors failed and the bypass pipe collapsed. Hot cyclohexane was emitted with flash vaporization and massive entrainment. Two distinct clouds were observed: a larger elevated cloud and a base cloud.

A minor explosion took place in the control room some 10–25 seconds after the release. When the base cloud reached the hot hydrogen unit, part of it was carried up by the thermal draft and ignited by the open burners at the top. This occurred some 22 seconds before the explosion of the elevated cloud. Flames were seen moving back to the escape point from the hydrogen plant and control room area and probably caused the elevated cloud to ignite some 54 seconds after the escape started. The main aerial explosion then occurred, followed by a major fire with fire-storm characteristics. For 20 minutes the fire raged over an area of 180 × 250 m with flames over 100 m in height. At the time of deflagration it was believed that the large aerial cloud contained about 45 tonnes of cyclohexane. Ninety percent of buildings on the site suffered damage, with blast being the primary factor. Fire extended the damage where the blast breached the containment of flammable inventories. The incident killed 28 people, all on the plant site. Over 400 people received treatment for injuries.

There has been much argument about the cause of failure of the bellows and much criticism of the way the bellows were installed. However, inventory levels were high. Each reactor had a capacity of 27 tonnes, which could empty in 10 minutes. The total process inventory was 400 tonnes of cyclohexane and cyclohexanone. Furthermore the pump rate through the reactors was large due to the low conversion in the reactors. This meant that a 10-minute flow corresponded to a throughput of 43 tonnes. This figure is closest to the estimated size of the cloud based on general evaluation of the explosion.

11.2.3 Seveso, Italy, July 1976

On Saturday 10 July 1976, an explosion occurred in a TCP (2,4,5-trichlorophenol) reactor belonging to the ICMESA chemical plant on the outskirts of Meda, a small town about 20 km north of Milan, Italy. A toxic cloud containing TCDD (2,3,7,8-tetrachlorodibenzo-p-dioxin), then widely believed to be one of the most toxic man-made chemicals, was accidentally released into the atmosphere. It was an exothermic decomposition that led to the release of dioxin-containing material to the atmosphere. The chemical reaction to produce the material had been performed earlier. Subsequently a number of plant activities had been carried out contrary to operating procedures. Operations only distilled off 15% instead of 50% of the total charge of ethylene, operators did not add water to cool the reaction mixture, and did not remain with the unit until the target cooling temperature was reached. It is considered bad practice to discharge a liquid directly to the atmosphere but in this case bout 2 kg of dioxin was discharged. Also, there has been justifiable criticism that the emergency plan was implemented slowly. An area of some 2 square miles was declared contaminated, a figure that was later increased by a factor of five. The release killed off large areas of vegetation and about 3300 animals were killed by the dioxin, while a further 70,000 were slaughtered to prevent dioxin from entering the food chain. People in the affected areas suffered from skin infections and other symptoms. At least 250 cases of the infection were identified, some 600 people were evacuated and the land was later decontaminated. Subsequent effects of the dioxin causing deaths in the long-term are still debated. The disaster led to the Seveso Directive (legislation), which was issued by the European Community and imposed much harsher industrial regulations.

11.2.4 Los Alfaques, Spain, July 1978

In this disaster on a Spanish camp-site, an articulated tank car carrying over 23 tonnes of liquefied petroleum developed a leak. The driver stopped beside a camp-site frequented by holidaymakers. Gas was released into the ground and ignited, probably in the camp-site disco. The tanker was engulfed in flames and this may have been followed by a BLEVE. Others have suggested a flash fire and a gas explosion. Certainly the people in the vicinity, largely holidaymakers and tourists, were singularly ill-protected against heat radiation; many were sunbathing at the time of the accident and large numbers of them were photographed watching a pall of smoke rising from the tanker, which was hidden by light trees. More than 200 people died on-site and a similar number succumbed to injuries, as they were unaware that they should have been escaping from the area rather than spectating. The tank car had been overloaded, it had no relief valve, and the steel tank was deteriorated, having been used for the transport of ammonia. Hence a crack developed. The route selected by the driver was the coastal road and it is presumed that no advice was given on what emergency procedures to follow in the event of a leak.

11.2.5 Mexico City, Mexico, November 1984

At approximately 05.35h on 19 November 1984, a major fire and a series of catastrophic explosions occurred at the government owned and operated PEMEX LPG Terminal at San Juan Ixhuatepec, Mexico City.

Some 11,000 m³ of liquefied petroleum gas (LPG) was stored in six 1600 m³ spheres and 48 horizontal cylindrical bullets, all in close proximity. The legs of the spheres were not fireproofed. It is believed that no fixed water sprays or deluge systems were fitted to the tanks. A leak of LPG from an unknown source formed a vapor cloud that was ignited by a pant flare. The storage area was bunded into 13 separate areas by walls of about 1 m high. A fierce fire developed, engulfing the spheres, which went up one after the other in a series of BLEVEs. Nine explosions were recorded. The accident is, to date, the most catastrophic accident in history involving domino effects. The series of LPG explosions at the distribution center resulted in 542 fatalities, and over 7000 people were injured. Some 200,000 people were evacuated. The fireballs were up to 300 m in diameter and lasted as long as 20 seconds. Rain consisting of liquid droplets of cooled LPG fell over the housing area covering people and property, and were consequently set alight by the heat from the fireballs. Since the construction of the plant some 100,000 people had settled in crowded housing on the valley floor and slopes. The local housing was mainly single storey and built of brick supported by concrete pillars. LPG was used for heating and cooking and each household had its own small bottles. Some 2000 houses at 300 m were destroyed and 1800 were badly damaged. Windows were broken at 600 m and debris missiles were thrown a considerable distance. One cylinder was thrown 1200 m. The emergency plan functioned well considering the circumstances. This accident is also called "the forgotten accident" due to another, even much worse, disaster that happened approximately a month after this accident: the Bhopal catastrophe (see next section).

11.2.6 Bhopal, India, December 1984

Ingress of water (because of pipe washing) in tank E610 initiated a runaway reaction that caused the release of some 25 tonnes of methyl isocyanate and probably hydrogen cyanide, effecting 3787 fatalities and 200,000 injuries instantaneously. In total, the death toll is estimated to have risen to a number as high as 20,000 people, which means that the Bhopal accident is considered to be the worst disaster ever to take place within the chemical industry. The cause was initially claimed by the company to have been sabotage. This was later denied by official investigators; and the incident undoubtedly reached the proportions it did because of operating instrumentation, the safety interlock systems and mitigating systems were inaccurate, inoperable or undersized. The standard of maintenance was appalling and the chemical plant should not have been operating under such conditions. The emergency plan was extremely poor with negligible communication to the public.

The material that leaked was an intermediate product that need not have been stored in such quantities. Alternative routes might have been adopted. As a result of the incident, Union Carbide, at that time one of the biggest chemical companies in the world, the owner of the plant, saw share ratings plummet on the US stock market, revealing how a major incident can also be a true financial disaster for any company.

11.2.7 Chernobyl, Ukraine, April 1986

An experiment was carried out to investigate whether or not a nuclear reactor could develop enough power to keep auxiliary equipment running while shutting down. The plant was operated below the power output at which a reactor remains stable. The design of the reactor made it liable to a positive void coefficient at power settings below 20% of maximum. At one point the power dipped below 1% of maximum and slowly stabilized at 7% of maximum. Operators and engineers continued to improvise by gradually removing rods. The plant went "super-prompt-critical" and an explosion followed. When the temperature increased it started to rise 100- fold in 1 second. The fire and radiation release caused many deaths. Exact numbers are unclear, but about 2 million people are believed to have been affected. The incident caused the permanent evacuation of 600,000 people and vast contamination.

11.2.8 Piper Alpha, North Sea, July 1988

A condensate pump on an oil rig tripped. The duty of the condensate system exceeded the initial design and such problems were not uncommon. Staff started up the spare, which had earlier been shut down for maintenance and during which the pressure relief cap had been removed and replaced by a cap that was not leak-proof. Clearly there were failures in communication of information during the shift changeover on the evening of the incident. Gas escaped from the cap and ignited. The resulting explosion destroyed the fire control and communication systems and demolished a firewall. The incoming gas pipeline was ruptured upstream of the emergency isolation valve and the gas burned fiercely as it does in a blowtorch. A fireball engulfed the platform. The adjacent rigs continued to feed gas and oil to Piper Alpha for over an hour. Other pipelines ruptured, intensifying the fire, and eventually most of the platform toppled into the water. The platform controller had tried to enact the emergency plan, which involved mustering in the galley followed by evacuation by helicopter. However, the explosions made escape by helicopter impossible. Some survivors escaped by jumping into the sea from a height of up to 50 m. 167 oil workers were killed, the platform was totally destroyed and UK hydrocarbon production dropped temporarily by 11%. Most of the fatalities were caused by smoke inhalation in the galley or accommodation areas.

11.2.9 Pasadena, Texas, USA, October 1989

Early afternoon on 23 October 1989 Phillips' 66 chemical complex at Pasadena, near Houston (USA), experienced a chemical release on the polyethylene plant. A flammable vapor cloud formed and subsequently ignited resulting in a massive vapor cloud explosion. Following this initial explosion there was a series of further explosions and fires.

The consequences of the explosions resulted in 23 fatalities and 314 people were injured. Extensive damage to the plant facilities occurred.

The day before the incident, scheduled maintenance work had begun to clear three of the six settling legs on a reactor. A specialist maintenance contractor was employed to carry out the work. A procedure was in place to isolate the leg to be worked on. During the clearing of the No. 2 settling leg, a part of the plug remained lodged in the pipework. A member of the team went to the control room to seek assistance. Shortly afterwards the release occurred. Approximately 2 minutes later the vapor cloud ignited.

11.2.10 Enschede, The Netherlands, May 2000

On 13 May 2000, a devastating explosion at a fireworks depot ripped through a residential district in the eastern Dutch town, leaving 22 people dead and 2000 families homeless. A small fire at the factory triggered several massive explosions. In this incident, several important necessary safety precautions had not been observed. The fireworks were not stored properly. They had been put in sea containers, offering insufficient delay in fire inhibition. To exacerbate matters, the classification on the boxes of fireworks was incorrect, leading the authorities and fire brigade to believe that they were dealing with consumer fireworks instead of professional explosives. Statistics about heat radiation require consumer fireworks to be at a distance of 20 m to inhabited buildings. Professional fireworks, however, need to be hundreds of meters away from residential areas.

11.2.11 Toulouse, France, September 2001

On 21 September 2001, an explosion in Shed 221 of the AZF (Azote de France) plant killed 31 people and injured 2442 people. The catastrophe cost the French government 228 million euros and TotalFinaElf, owner of AZF, more than 2 billion euros.

Although there remain some uncertainties and unsolved questions about the disaster, it has been assumed that the explosion was caused by a human handling error. A worker from a subcontracted company mistook a 500-kg sack of a chlorine compound (dichloroisocyanuric acid) for nitrate granules and poured it onto the stock of ammonium nitrate in Shed 221 15 minutes before the explosion. The mixture is said to have produced trichloroamine, an unstable gas that explodes at normal temperatures.

11.2.12 Ath, Belgium, July 2004

A huge gas explosion occurred at about 09.00 h on 30 July 2004 in the small Belgian town of Ghislenghien just outside Ath, 40 km south of Brussels. It sent a wall of flame into the air, triggering a chain of explosions. A leak was reported on the pipeline, which runs from the Belgian port of Zeebrugge into northern France, 37 minutes before the explosion. Firefighters attempting to establish a security perimeter around the site were among those killed when the explosions destroyed two factories in the industrial park. The blast was heard several miles away. It melted or burned everything within a 400-m radius and left a large crater between the two factories. Bodies and debris were thrown 100 m into surrounding fields. Twenty-four people died in the accident and 120 people were injured, half of them seriously.

11.2.13 Houston, Texas, USA, March 2005

At approximately 13.20h on 23 March 2005 a series of explosions occurred at the BP Texas City refinery during the restarting of a hydrocarbon isomerization unit. Fifteen workers were killed and about 170 others were injured. Many of the victims were in or around work trailers located near an atmospheric vent stack. Investigators reported that the explosions occurred when a 170-feet distillation tower flooded with hydrocarbons and was overpressurized, causing a geyser-like release from the vent stack. As a result of this mistake, a mixture of liquid and gas flowed out of the gas line at the top of the column, travelled through emergency overflow piping, and was discharged from a tall vent that was located hundreds of feet away from the distillation column. A vapor cloud accumulated at or near ground level. The cloud was further ignited by a vehicle that had been left in the area with its engine idling. Finally, a number of mobile offices that had been located far too close to the plant were destroyed by the explosion, killing and injuring their occupants.

11.2.14 St Louis, Missouri, USA, June 2005

St. Louis was experiencing a heat wave, with bright sunlight and temperatures exceeding 35°C on June 24 2005. At Praxair, a gas repackaging plant, operations proceeded normally during the morning and early afternoon; however, in the afternoon a security camera video from the facility shows the release and ignition of gas from a cylinder in the propylene return area. As workers and customers evacuated, the fire spread to adjacent cylinders. The video shows nearby cylinders igniting in the first minute. At 2 minutes, cylinders began exploding, flying into other areas of the facility, and spreading the fire. After 4 minutes, the fire covered most of the facility's flammable gas cylinder area and explosions were frequent. Fire swept through thousands of flammable gas cylinders and dozens of exploding cylinders were launched into the surrounding community and struck nearby homes, buildings, and cars, causing extensive damage and several small fires.

11.2.15 Buncefield, UK, December 2005

From around 18.50 h on Saturday 10 December 2005 a delivery of unleaded petrol was being pumped down the T/K pipeline from Coryton Oil Refinery into tank 912 (situated within bund "A"). The automatic tank gauging system, which records and displays the level in the tanks, had stopped indicating any rise in tank 912 fuel level from around 03.00 h on Sunday 11 December. At about 05.40 h on Sunday morning, tank 912 started to overflow from the top. The safety systems that were designed to shut off the supply of petrol to prevent overfilling, failed to operate. Petrol cascaded down the side of the tank, collecting in bund A. As overfilling continued, a vapor cloud that was formed by the mixture of petrol and air, flowed over the bund wall, dispersed and flowed off the site and towards the Maylands industrial estate. Up to 190 tonnes of petrol escaped from the tank, about 10% of which turned to vapor that mixed with the cold air, eventually reaching concentrations capable of supporting combustion. The release of fuel and vapor is considered to be the initiating event for the explosion and subsequent fire.

At 06.01 h on 11 December 2005, the first of a series of explosions took place. The main explosion was massive and appears to have been centered on the Maylands Estate car parks. These explosions caused a huge fire, which engulfed more than 20 large storage tanks over a large part of the Buncefield depot. The fire burned for 5 days and a plume of black smoke from the burning fuel rose high into the atmosphere. There were only 43 injuries, but the human death toll could have been very large if the accident had happened during a regular day of the week instead of during the night at the weekend. The damage was estimated at several billion euros.

11.2.16 Port Wenworth, Georgia, USA, February 2008

On February 7, 2008, a huge explosion and fire occurred at the Imperial Sugar refinery northwest of Savannah, Georgia, causing 14 deaths and injuring 38 others, including 14 with serious and life-threatening burns. The explosion was fuelled by massive accumulations of combustible sugar dust throughout the packaging building. The investigation report issued by US Chemical Safety and Hazard Investigation Board concluded that the initial blast ignited inside a conveyor belt that carried sugar from the refinery's silos to a vast packaging plant where workers bagged sugar under the Dixie Crystals brand.

11.2.17 Deepwater Horizon, Gulf of Mexico, April 2010

A fire aboard the oil rig Deepwater Horizon started at 9.56 p.m. on 20 April 2010. At the time, there were 126 crew on board. Suddenly, two strong vibrations were felt by the employees. A bubble of methane gas escaped from the well and shot up the drill column, expanding quickly as it burst through several seals and barriers before exploding. The event was basically a blowout, and indeed a number of significant problems have been identified with the blowout preventer. Survivors described the incident as a sudden explosion, which gave them less than five minutes to escape as the alarm went off. The explosion was followed by a fire that engulfed the platform. After burning for more than a day, Deepwater Horizon sank on 22 April 2010 at approximately 10.21 h. As a result of the accident, 11 workers died, and approximately 5 million barrels of oil (790,000 m^3) were spilled into the Gulf of Mexico. Besides the huge immediate consequences for all life at sea, in October 2011 dolphins and whales continued to die at twice the normal rate. In April 2012, 2 years after the accident, scientists reported finding alarming numbers of mutated crab, shrimp and fish they believe to be the result of chemicals released during the oil spill.

11.2.18 Fukushima, Japan, March 2011

Following a major earthquake, a 15-m tsunami disabled the power supply and cooling of three Fukushima Daiichi reactors, causing a nuclear accident on 11 March 2011. All three cores largely melted in the first 3 days. High radioactive releases were measured. After 2 weeks the three reactors were stable with water addition but no proper heat sink for the removal of decay heat from the fuel. By July 2011 they were being cooled with recycled water from a new treatment plant. Reactor temperatures had fallen to below 80°C at the end of October 2011, and an official "cold shutdown condition" was announced in mid-December 2011. Apart from cooling, the basic on-going task is to prevent release of radioactive materials, particularly in contaminated water leaked from the three units. There have been no deaths or cases of radiation sickness reported from the NaTech nuclear accident as yet, but over 100,000 people have had to be evacuated from their homes.

11.3 Learning from accidents

Many disasters have occurred because organizations have ignored the warning signs of precursor incidents or have failed to learn from the lessons of the past. Normal accident theory suggests that disasters are the unwanted, but inevitable output of complex socio-technical systems, while high reliability theory sees disasters as preventable. by certain characteristics or response systems of the organization (see Chapter 3, Section 3.10.2). Industrial accidents and hazards have become the order of the day, with new technologies evolving every day and few people knowing how to use them. Disasters have at least one thing in common: the inability of the organization involved to effectively synthesize and share the information from separate "precursor" incidents with the relevant people across the organization so that appropriate action could be taken to reduce the risk of disaster. Kletz (6) reports several examples in the chemical industry of the same accident occurring multiple times in the same organization. We could then suppose that it is not natural for organizations to learn from safety incidents. Even if ad hoc learning is occurring, it is not enough.

It is then essential that incidents and accidents are properly reported. The investigation of any accident will never progress unless it is first properly reported within an organization. A formal policy requiring the consistent and adequate reporting of all incidents and accidents is one of the most important principles of any accident investigation program. Much of what is known today about accident prevention and loss control was, in fact, learned from loss incidents that were properly reported and investigated.

The accident investigation should be objective and complete. It is all too common that accident investigations fail to identify the reasons that an accident occurred because the investigator focused on assigning blame rather than determining the underlying causes. This is the reason that people are reluctant to report any event that might reflect unfavorably on their own performance or that of their department. However, it is obvious that without complete reporting of incidents, near-misses, accidents, losses, etc., and followed by a comprehensive and "honest" investigation into the causes, an organization and its management will never know the extent and nature of the conditions that no doubt will have downgraded the efficiency of the company.

To efficiently investigate accidents, management should determine appropriate parameters within which the investigation is to be carried out. Parameters may include the types of occurrences that will require reporting and investigating, to what extent the investigation has to be conducted, how incidents and accidents shall be reported and what information should be given, what use shall be made of the information reported, etc.

Vincoli (7) indicates that the fact that an accident occurs is a strong indication that a bad and/or erroneous decision was made by management within the organization somewhere. The nature of accident investigation requires an analysis of all possible reasons such decisions have been made. It is obvious, also from the examples above, that faulty communications, lack of adequate information, improper training, and improper behavior are examples of accident causal factors that are often repeated.

There are nine essential elements of a successful accident investigation program:

1. Consistent reporting to management.
2. Interview of witnesses and examination of evidence at scene.
3. Determination of immediate causal factors.
4. Study of all evidence and formulation of interpretations.
5. Determination of basis or root causal factors.
6. Reconstruction of accident (if required).
7. Analysis of basic causal factors and management involvement.
8. Implementation of corrective actions.
9. Follow-up of planned and implemented actions.

Specific plans should be developed to organize and manage the carrying out of an accident investigation in anticipation of any possible future incident or accident. Adequate planning ensures that sufficient resources will be available to guarantee a proper investigation if an incident or accident should occur. Individual or team investigation, team member composition depending on the characteristics of the incident or accident, coordination requirements, collaboration specificities between investigators, required background information and data, exchange of information, sharing ideas and brainstorming sessions, training, expertise, know-how, information, resources, etc., should all be thought of in advance. Hence, preparation is the key to the accurate determination of incident and accident causal factors, in the shortest possible time and with the lowest expenditure of resources.

What does it take to learn?

- Opportunity (to learn) – Learning situations (cases) must be frequent enough for a learning practice to develop.
- Comparable/similar – Learning situations must have enough in common to allow for generalization.
- Opportunity (to verify): It must be possible to verify that the learning was "correct" (feedback).

The purpose of learning (from accidents, etc.) is to change behavior so that certain outcomes become more likely and other outcomes less likely. According to Kletz (2001), there are several *myths on accident investigation* to note [see (8) for supplementary information] and to develop with your own culture, education and goal. The myths are:

1. Most accidents are caused by human error. (The authors of this book would suggest to use/interpret "most" into "all", to avoid possible misunderstandings. Another possibility is to explicitly indicate that with "accidents", Kletz means "major accidents".)
2. Natural accidents are "Acts of God" and are unavoidable.
3. The purpose of accident reports is to establish blame.
4. If someone's action has resulted in an accident, the extent of the damage or injuries is a measure of his or her guilt.
5. We have reached the limit of what can be done by engineering to reduce accidents. We must now concentrate on human factors.
6. Most industrial accidents occur because managers put costs and output before safety.
7. Most industrial accidents occur because workers fail to follow instructions.
8. Most industrial accidents occur because workers are unaware of the hazards.
9. Two competent and experienced investigating teams will find the same causes for an accident.
10. When senior people do not know what is going on this is because of the failures of their subordinates to tell them.
11. When everything has run smoothly for a long time, and accidents are few, we know we have got everything under control.

Once you have debated these myths, these comments could be raised:

- As individuals, we do learn from experience of accidents and rarely let the same one happen again, but when we leave a company we take our memories with us. We need to find ways of passing on our knowledge and building a "company memory".
- The next best thing to experience is not reading about an accident, or listening to a talk on it, but discussing it.
- When we investigate an accident, and particularly if we are involved in some way, it is hard for us to be dispassionate and our mindsets and self-interests can easily take over. Without making a conscious decision to do so, we look for shortcomings in other departments rather than our own and tend to blame those below us rather than those above us.

11.4 Conclusions

Accidents happened. Accidents happen. Accidents will happen. It is obvious that a zero risk or zero-accident organization does not exist. Industrial activities go hand-in-hand with risks and regretfully with incidents and accidents. This does not mean however that nothing can be done to prevent most incidents and accidents, or to prevent the consequences of accidents aggravating. Accidents will happen within organizations, but the number of accidents should be very low, and the consequences of accidents should be none to minor, when applying engineering risk management and its principles, methods, concepts, etc. explained and discussed in this book. We need especially to learn from minor and major accidents and insert the lessons learned of any disaster worldwide into the DNA and the memory of all organizations. Hence, on the bright side, the vast majority of possible incidents and accidents never happen, thanks to adequate engineering risk management. On the dark side, disasters keep on taking place somewhere worldwide, and they really are preventable, either by preventing the initiating causes of the disaster, or by preventing the accident's consequences from becoming disastrous.

References

1. Kahneman, D., Tversky, A. (1979) Prospect theory: An analysis of decisions under risk. Econometrica 47:263–291.
2. Sutherland, V., Making, P., Cox, Ch. (2000) The Management of Safety. London, UK: Sage Publications.
3. Wells, G. (1997) Major Hazards and their Management. Rugby, UK: Institution of Chemical Engineers.
4. Reniers, G. (2006) Shaping an integrated cluster safety culture in the chemical process industry, PhD dissertation. Antwerpen, Belgium: University of Antwerp.
5. Paltrinieri, N. (2012) Development of advanced tools and methods for the assessment and management of risk due to atypical major accident scenarios, PhD dissertation. Bologna, Italy: University of Bologna.
6. Kletz, T.A. (1993) Lessons From Disaster – How Organizations have No Memory and Accidents Recur. Houston, TX: Gulf Publishing Co.
7. Vincoli, J.W. (1994) Basic Guide to Accident Investigation and Loss Control. New York: John Wiley & Sons.
8. Kletz, T.A. (2001) Learning from Accidents. Oxford, UK: Butterworth-Heinemann.

12 Concluding remarks

In any organization, practices regarding engineering risk management are without a doubt what make the difference in safety performance, and by extent, in business performance. When it comes to making a difference, there is no goal more important than ensuring that risks (positive and negative) are well-managed and that every employee goes home safe (and happy) every day. The managers of long-term successful organizations are indeed well aware that decreasing negative uncertainties and risks and guaranteeing safety for their employees makes the difference between their companies and less successful organizations. Hence, engineering risk management is very important.

Risk is understood as an uncertain consequence of an event or an activity with respect to something that human's value. In its simplest form, risks refer to a combination of two components: the likelihood or chance of potential consequences and the severity of consequences of human activities, natural events or a combination of both. Such consequences can be positive or negative, depending on the values that people associate with them. Often engineers concentrate their efforts on (predominantly negatively evaluated) risks that lead to physical consequences in terms of human life, health, and the natural and built environment. It also addresses impacts on financial assets, economic investments, social institutions, cultural heritage or psychological well-being, as long as these impacts are associated with the physical consequences. In addition to the strength and likelihood of these consequences, this book emphasizes the distribution of risks within and across organizations and, over time, space and people. In particular, the timescale of appearance of adverse effects is very important and links risk management to sustainable development (delayed effects).

We distinguish risks from hazards. Hazards describe the potential for harm or other consequences of interest. These potentials may never even materialize if, for example, people are not exposed to the hazards or if the targets are made resilient against the hazardous effect (such as immunization). In conceptual terms, hazards characterize the inherent properties of the risk agent and related processes, whereas risks describe the potential effects that these hazards are likely to cause on specific targets such as buildings, ecosystems or human organisms and their related probabilities.

There are a lot of books and other literature that discuss how to conduct non-financial risk management. However, these works only focus on the purely technical aspects of what risk management is all about, which is only a part of the story. Or they discuss it from a non-technological viewpoint and they only very briefly discuss technical matters of risk management. This book tries to provide an inclusive overview and treats risk management from an engineering standpoint, thereby including technological ("engineering") topics (such as what risk assessment techniques are available and how to carry them out, how to conduct an event analysis, etc.), as well as non-technological ("management") topics (such as risk management concepts, crisis management, risk governance, economic issues related to risks, etc.). There is a reason why it is so difficult to present and write an inclusive book on risk management: risk management is confronted with three major challenges that can be best described using the terms "complexity", "uncertainty" and "ambiguity". These three challenges are not related to the intrinsic characteristics of hazards or risks themselves but to the state and quality of knowledge available about both hazards and risks.

- Complexity refers to the difficulty of identifying and quantifying causal links between a multitude of potential causal agents and specific observed effects.
- Uncertainty is different from complexity but often results from an incomplete or inadequate reduction of complexity in modeling cause-effect chains. It might be defined as a state of knowledge in which, although the factors influencing the issues are identified, the likelihood of any adverse effect or the effects themselves cannot be precisely described.
- Ambiguity – Whereas uncertainty refers to a lack of clarity over the scientific or technical basis for decision-making, ambiguity is a result of divergent or contested perspectives on the justification, severity or wider "meanings" associated with a given threat.

In economics, the "Trias Economica" exists, implying that a healthy economy needs three partners: financial institutions (banks), insurance companies, and all other companies. If one of these partners falls away or does not perform its function in the economy, the economy will heavily suffer. In risk management, a parallel can be drawn: the "Trias Risico" (Risk Trias) is proposed, indicating that to exist, a risk needs three factors: one or more hazards, exposure to the hazard, and possible loss. If one of these factors is eliminated, the risk ceases to exist. Risk management therefore is concerned with influencing/decreasing the importance, in any way (this can be extremely technical to procedural, to purely human factor, communication, etc.), of one or a combination of these three factors. This seems to be very simple, but it is not. Every individual factor and all its influencing parameters can be very complicated or complex to deal with, showing high levels of uncertainty, high levels of ambiguity, high requirements of knowledge and collaboration, (hidden) relationships between the parameters, etc. Because of this complexity, a systemic engineering approach is absolutely needed, besides an analytic engineering approach.

In the past "risk management" was sometimes a synonym for "insurance management". Luckily, this is not at all the case anymore. Nowadays, organizations are well aware that non-financial risk management is a very important domain that indirectly leads to a large amount of financial gains. However, the domain is so wide that there seems to be no consensus about what are the best approaches and models to use, or which are the most efficient analysis techniques, etc. This book therefore fills a gap in presenting in an easy and legible way, the knowledge and know-how about risk management in order to be successful. Indeed, applying the various ideas, concepts and techniques provided and elaborated in this book, leads to more effective and more efficient risk decision-making and to better or to optimal results.

Nonetheless, we have to accept that the most debated part of handling risks refers to the process of delineating and justifying a judgment about the tolerability or acceptability of a given risk. The term "tolerable" refers to an activity that is seen as worth pursuing (for the benefit it carries) yet it requires additional efforts for risk reduction within reasonable limits. Whereas, the term "acceptable" refers to an activity where the remaining risks are so low that additional efforts for risk reduction are not seen as necessary.

In the end, which risk could be accepted, tolerated or just taken, will largely be driven by other factors than purely technical ones. Society, culture, ethics, economy, regulations and business appetite will largely influence the decision-makers. However, to take the best decision, it should be based on solid analysis and evaluation in order that the risk taken – because "zero risk" or "absolute safety" is a myth – is decreased as much as sustainably possible to a level that not only the business is taking, but also the community as a whole. A major challenge for risk managers and engineers resides in the best supply of information, and in a broad sense communication, to decision- or policy-makers in order to achieve the best risk governance for a better world.

In popular imagination, rocket science is the totemic example of scientific complexity. Risk management is indeed no rocket science. It is much more complicated than rocket science! After all, there are not ten possible approaches to send a rocket to the moon or to Mars. There are, nonetheless, tens of thousands of possible risks, and easily hundreds of possible approaches to manage them, leading to as many different possible outcomes and realities.

Einstein mentioned that if he had 1 hour to solve a problem, he would spend 45 minutes in understanding and analyzing the problem, 10 minutes to perform a critical review and finally use the last 5 minutes to solve it. Transposing this to risk management leads to exactly the same concept, except that the last 5 minutes will be taken for the decision-making process.

Index

acceptability, 29, 136
acceptable, 237
acceptance, 170
acceptation, 150
accident, 12, 43, 48, 53, 138, 147, 152, 173, 177, 180, 206, 207, 241
accident analysis, 177, 185
Accident costs, 204
accident investigation, 269
accident investigation program, 269
accident pyramid, 37
accident ratio, 37
Accident reporting, 82
accident reports, 177
accident sequence, 174, 175
Achievable, 69
action plan, 98, 103, 184
actions, 196
acute, 251
advances, 251
Agricola, 2
ALARA, 61, 149
ALARP, 49, 149
allocation, 204, 215
alternative Risk, 147
ambiguity, 274
analogy, 206, 253
analysis, 101, 179
analytic, 3
analytical, 140
Anatomy, 43
Anglo Saxon model, 220
Anticipation, 190
anxiety, 190
approach, 2
arcs, 140
artifacts, 62
asphyxia, 250
Asphyxiation, 247
assessment, 225, 273
Ath, 266
attack, 46
audit, 195, 244, 246
auditing, 224

authorities, 192, 261
availability, 81, 114
avoidance, 170
awareness, 62
axes, 135
AZF, 156, 266

banding, 255
bands, 135
barrier, 39, 53, 130
basic process control systems, 42
bathtub curve, 81
Bayes, 142
Bayesian Network, 140
BCP, 194
behavior, 22, 174, 199, 226, 236
benefit, 204, 207, 210, 252
Bhopal, 155, 264
biopersistence, 254
bi-pyramid, 37
Bird, 207
black swan accidents, 12
blast, 248
BLEVE, 261
block, 126
BN, 143
board, 220
Boolean algebra, 123
bottom-up, 116
BPCS, 42
BP Texas City, 266
brainstorming, 150
branch, 126
break-even, 210
Buncefield, 267
business, 195
business appetite, 275
business continuity plan, 194
business recovery, 189

caprolactam, 261
carcinogenic, 244
cataclysm, 188
catastrophe, 197

categorize, 130
Causal, 176
causal factor, 143, 177
causal tree, 177, 182, 183
cause, 76, 87, 124, 140, 181, 182, 186
cause-consequence, 138
cause-consequence-analysis, 127
CCA, 127
chain, 139
chance, 4, 273
change, 251
checklist, 103
chemicals, 245
Chernobyl, 264
chronic, 251
chronological overview, 259
chronology, 197
circumstances, 173, 178
classification, 255
climate, 57
cluster, 221, 237
CMR, 244, 255
cognitive, 237
collective mindfulness, 79
collective protection, 151
combinations, 130
communicating vessels, 75
communication, 150, 185, 190, 243, 253
communication plan, 252
communication strategy, 199
communicator, 199
Company management, 230
company's H&S standards, 209
complex, 6, 115
complexity, 6, 253, 274
component, 115, 117
conception of risk, 33
Conditional Probability Table, 140
conflict, 200
connectors, 182
consequence, 2, 16, 87, 103, 126, 149, 251, 273
contamination, 264
continuity, 189
continuity plan, 191
continuous, 63
continuous improvement, 62, 231
Continuous risk management, 8
contours, 136
Contractor, 224
control, 38, 114, 224, 243

control banding, 255
coordinator, 243
corporate, 59
Corporate Governance, 219
corrective, 106
corrective measures, 111, 186
cost, 103, 122, 204, 212
cost-benefit, 170, 206
Cost-benefit analysis, 213
Cost-effectiveness analysis, 212
Cost evolution, 96
cost point, 209
countermeasures, 175
coupled systems, 77
coupling, 77
CPT, 140
crisis, 187
crisis communication, 190, 199
crisis management, 187
criteria, 50, 136, 138
criticality, 16, 116, 119
cryogens, 246
cube, 137
culture, 220, 252, 274
Cut-Sets, 123

damage, 244
dangerous goods, 224
database, 245
Data collection, 176
DBN, 143
decision, 6, 216, 238
decision tree, 254
decision-makers, 155, 203, 229
decision-making, 97, 144, 188, 203, 238, 274, 275
decomposition, 117
deductive, 88, 123
Deepwater, 1
Deepwater Horizon, 268
deficiencies, 236
deficits, 236
degradation, 150, 156, 188
degree of safety, 209
delegate, 243
Deming, 62
Deming cycle, 63
description, 108
design stage, 96
detectability, 119
detection, 114, 158

Deterministic, 90
devastation, 12
deviations, 104, 108, 112
Diagnosis, 238
diagrams, 126
dioxin, 263
direct, 204
directive, 151, 243
disaster, 150, 187, 189, 194, 200, 264
disincentives, 148
disruption, 8, 79, 154, 191, 200
Documentation, 224
domino, 41, 175
domino model, 41
domino theory, 41, 175
dynamic, 122, 251
Dynamic model, 54

ecology, 154
economic, 203, 216, 252
economy, 154, 274
education, 243
effect, 140, 274
efficiency, 252
EHS, 1
Einstein, 275
elementary, 117
elimination, 147
emergency, 192, 194, 243
emergency plan, 157
emerging risk, 253
emerging technology, 251
employee, 224
empowerment, 147
engineering, 1, 91, 274
Engineering risk, 24
engineering risk management, 3, 29
ENP, 254
Enschede, 265
environment, 45, 80, 259
escape, 148
estimates, 214
ETA, 126, 127
ethics, 274
event, 12, 181, 246
event analysis, 173
Event tree analysis, 123, 126
evidence, 177
excellence, 60
execution, 79
expert, 229, 252

Explosion, 242
exposure, 251, 254, 255

facilitator, 112
factors, 173
facts, 178, 181
failure, 80, 103, 104, 112, 114, 122, 123,
    127, 136, 158, 236, 256
Failure Mode and Effects Analysis, 114
failure of risk, 4
FAR, 50
Fatal Accident Rate, 50
fatality, 136, 154
fatality risks, 49
fault, 123, 173
fault diagram, 123
fault trees, 130
Fear, 2
feared event, 124
Feyzin, 261
final event, 184
Financial, 210
firewalls, 156
flammable, 247
Flixborough, 107, 155, 261
FMEA, 114
FMECA, 114
FN-curve, 49, 136
follow-up, 185
forecast, 26, 98, 198
fractures, 248
framework, 221, 230
frequency, 136, 138, 139
frequency rate, 83
frostbite, 247
FTA, 123, 127
Fukushima, 268
function, 111
function loss, 118
functional diagram, 115

gates, 123
Ghislenghien, 266
GHS, 152
governance, 188, 190, 226, 236,
    252, 275
government, 153
governmental, 220
Group meetings, 223
guards, 46
guidance, 123, 254

guide keywords, 108
guidelines, 108
guidewords, 110

hazard, 5, 14, 164, 243, 250, 273
Hazard analysis, 224
Hazard identification, 103
Hazard mapping, 19
hazard portfolio, 20
HAZOP, 107
health, 154, 247
hierarchical, 115
High Reliability Organizations, 79
High Reliability Theory, 76
Hippocrates, 2
holistic, 8
housekeeping, 149
HRO, 79
HRT, 76
human, 136
human behavior, 190
human senses, 250
hurricane, 197
hypothermia, 247
hypothetical benefits, 204

identification, 33, 35, 101, 104
impact, 251, 253, 254
Imperial Chemical Industries, 107
implementation, 121
importance, 29
improvement, 36
incentives, 236
incident, 37, 43, 53, 147, 173, 177, 207, 241
income, 206
indications, 79
indicators, 64, 65, 73
indirect costs, 204
individual, 138
individual risk, 48
inductive, 88, 112, 114, 130
inductive-deductive, 130
industrial accidents, 259
information, 199, 243
initiating event, 126, 130, 136
innovation, 253
input, 108, 115, 123
inspection, 224
instability, 188, 190
institutional, 203, 220
insurance, 274

integrated management system, 35
integrated risk management, 35
interactions, 77, 166, 242
interconnected, 112
interface, 167
interpretation, 16, 173, 179
interview, 179, 180
inventory, 246
investigation, 177, 180, 181, 224, 269
investigators, 176
investment, 216
IPLs, 139
ISO, 64
ISO 9001, 223
ISO 14001, 223
ISO 31000, 7, 26
Iso-risk contours, 49
ISO Standard, 7
iterated expectations, 233
iteration, 96
iterative, 94

judgment, 130, 178, 179

Katrina, 197
key performance indicators, 236
keywords, 110, 114
knowledge, 5, 237, 251
Known Knowns, 14
Known Unknowns, 14

labeling, 250
lagging, 73
Layer, 138, 140, 233
Layer of Protection Analysis, 138
layers of protection, 42, 43
leadership, 60, 257
leading, 73
legislation, 259, 260
legislative, 154
level of comfort, 22
liability, 213
likelihood, 15, 123, 131, 147, 273
likelihood of occurrence, 16
liquefied petroleum gas, 263
liquid nitrogen, 250
litigation, 199
logic, 126
logical paths, 123
logic gates, 123
long term, 74

Long term vision, 76
loop, 63
Loop risk management, 222
LOPA, 138
Los Alfaques, 263
loss aversion, 216
losses, 48, 147
Lost Time Injury frequency rate, 82
Lost Time Injury Incidence Rate, 82
Lost Time Injury Severity Rate, 83
LPG, 263
LTIFR, 82
LTIIR, 83
LTISR, 83

magnetic, 246
maintainability, 81, 114
Maintenance, 166, 224
major accident, 259
major event, 187
Major hazard, 259
major industrial accidents, 260
Management, 243
management framework, 27
management of change, 242
management process, 24
management systems, 7, 220
managers, 203, 273
mapping, 243
Maslow, 2
matrix, 17, 99, 132, 254
Mayan pyramid, 37
Measurable, 69
measures, 256
media, 192
Medical Treatment Injury, 83
mesh, 112
metaphor, 75
methodology, 122
Mexico City, 263
MICE, 243
micro-economic, 210
mind-set, 9
mishap, 101
mistakes, 199
mitigate, 140
mitigation, 106, 126, 156, 181
monitoring, 224, 253
MSDS, 242
MTI, 83
multilinear, 174

multiple dimensions, 237
Murphy's Law, 191
mutagenic, 244
myths, 270
nanomaterials, 254
Nanotechnologies, 254
Nano safety, 254
NAT, 76
near-accident, 53
near-miss, 5, 53, 173, 174
negative, 14, 273
negative risk, 1, 16
net present value, 214
new technologies, 241
nodes, 140
non-financial, 273
Non-linear causalities, 76
non-occurring accidents, 204
Normal Accident Theory, 76
NPV, 214

objectives, 29, 36, 65
observations, 174
occupational, 211, 255
occurrence, 111, 123, 147
OHSAS, 64
OHSAS 18001, 7, 223
operating procedure, 104, 105
operational, 236, 238
operations, 194
opportunity, 10, 98, 199
organizational, 8, 26, 60, 80, 130, 166, 238
organizational culture, 56
outcome, 8
Outputs, 115

P2T, 38, 61
PABC-model, 22
parameter, 109
passive, 130
path, 124
pathogen, 175
pathways, 126
PDCA, 62, 231, 234
PDRC, 233
people, 62
perception, 21, 27, 33, 241, 251
performance, 72, 273
personal, 167
petrol, 267
PFD, 139

PHA, 101
Phillips, 265
physical, 52, 250
Physical risk model, 52
Piper Alpha, 264
Piping and Instrumentation
    diagram, 108
plan–do–check–act, 63, 221, 223, 243
Pliny, 2
policy, 57, 234
political, 220
positive, 273
potential losses, 163
powder, 254
practice, 12, 241
pragmatic, 256
Praxair, 266
precautionary, 253, 256
precautionary principle, 229
preconditions, 175
preliminary, 101, 103
Preliminary hazard analysis, 101
press release, 199
prevention, 17, 53, 130, 147, 177, 186, 211, 231, 255, 259
Prevention benefits, 208
prevention costs, 208
prevention plan, 151
preventive, 89, 148, 157
principal agent theory, 220
principles of risk management, 26
Proactive, 198
Probabilistic, 90
probability, 4, 33, 52, 96, 112, 124, 126, 140, 143, 216
Procedure, 61, 62, 122
procedure list, 103
process, 108, 109, 122
process description, 108
Process Flow Diagram, 108
Process Safety Management, 242
Product, 122
Production, 122
promotion, 150
propagation, 115
Proposal, 153
prospective, 89
protection, 17, 53, 130, 138, 156, 157
protective, 148
protocol, 190
psychology, 217, 231

public, 251, 259
public image, 189

QRA, 51, 135
qualitative, 93, 118
Quality, 195
quantification, 125, 130
quantitative, 93, 118, 127, 138
Quantitative Risk Assessment, 51, 135
questions, 103

Rapid Risk Ranking, 103
rays, 246
receiver, 199
recognition, 253
recommendations, 9, 107, 153, 176
records, 224
recovery, 148, 159, 189, 193
recurrence, 127, 173
reduction, 170
reduction/mitigation, 147
regulation, 152, 224, 236, 250, 275
relationship, 140, 192
Relevant, 69
reliability, 81, 114, 122, 126, 136
remedial, 148
remediation, 244
report, 154, 173
reprotoxic, 244
requirements, 225
resilience, 163, 170, 253
resources, 161
response, 159, 189, 192, 199, 253
responsibility, 169, 173
restoration, 193
retardant effect, 75
returns, 209
RGM, 231
rig, 265
Rijnland-model, 220
rings of protection, 46
risicare, 33
risk, 7, 252, 273
risk analysis, 33, 89, 94, 160
risk analysis techniques, 144
risk assessment, 15, 33, 87, 132, 230
risk attitude, 21, 133
risk averse, 22, 49, 259
risk averting, 226
risk calculation, 17
risk communication, 23

risk control, 164
risk displacement, 185
risk financing, 147
risk governance, 219, 234
Risk Governance Model, 231
risk issue, 238
risk management, 4, 33, 242, 257
risk management process, 23, 34
risk management set, 34
Risk management standard, 7
risk management systems, 7
risk managers, 216
risk mapping, 18
Risk Matrix, 130
risk mitigation, 4, 255
Risk perception, 21
risk priority number, 119
risk reduction, 147
risk society, 219
risk taking, 259
risk tolerance, 257
risk transfer, 147
Risk triangle, 231
Risk Trias, 11, 15
risk value, 16
risk/opportunity, 203
risk-addicted, 22
risk-paranoid, 22
risk-seeking, 22
risk-taking, 226
root cause, 175, 176, 185
routines, 80
Rumors, 173

safe, 107
safeguards, 42, 87, 138
safety, 65, 114, 154, 204, 244
safety barriers, 157, 188
safety culture, 57, 59
Safety economics, 210
safety engineering, 122
safety instrumented functions, 41
safety integrity level, 42, 44
safety layers, 42
safety management, 241
Safety Management System, 221
safety policy, 209
Safety promotion, 224
safety regulations, 259
safety rules, 224
safety standards, 225

Safety Training, 223
Safe work practices, 223
sandglass, 10, 11, 71
scenario, 126, 130, 136, 138, 190, 194
scenario building, 9, 226
Schweizerhalle, 151, 156
security, 65
segregation, 147
semi-quantitative, 93, 103, 107
sequence, 127
sequencing, 175
sequential events, 174
Service, 122
severity, 15, 16, 111
Seveso, 107, 151, 155, 262
short term, 74
SIF, 41
signals, 199
SIL, 42
simplifications, 79
SMART, 69
SMS, 221
social, 251
societal, 29, 50, 138, 154
societal risk, 49
society, 2, 274
socio-economic, 205
solutions, 185
source, 167
specialist, 256
Specific, 69
spills, 244
spokesperson, 199
Staffing, 208
stakeholder theory, 220
stakeholders, 27, 234, 253
standard, 107, 225
standpoint, 273
Static model, 54
STOP, 166, 169
storage, 246
strategic, 100, 167, 238
Strategic management, 62
strategic planning, 26
strategy, 7, 192, 215, 235, 251
Strengths, 10, 98
sub-dimensions, 66
substitution, 147, 168
subsystems, 115, 116
Sugar refinery, 267
surprise, 187

surveillance, 192, 252
survivability, 187
suspension, 254
sustainability, 122, 256
sustainable, 163, 234, 273
Swiss cheese model, 39
SWOT, 10, 35, 71, 98
system safety, 101
systematic, 107
systematic methods, 87
systemic, 3, 251, 274
systems, 126
systems thinking, 75, 229

tactic, 238
target, 52, 167, 188, 273
technical, 166, 273
technology, 62, 251
testimony, 180
Theorem, 142
thinking, 39
threat, 2, 10, 52, 62, 98, 135, 144, 166, 188
thresholds, 155
Time bound, 69
tolerability, 135
tolerable, 149
tolerance, 170
tools, 5
Top Event, 123
top-down, 116, 123
Total Recordable Frequency Injury Rate, 83
training, 243
transference, 170
tree, 124, 126, 130, 182
TRIFR, 83
triggers, 253
turnover, 241
tutorials, 243
type I, 37
type I accidents, 212

Type I events, 14
type I risks, 223
type II, 37, 223
type II accidents, 12, 212
Type II events, 14
type III accidents, 212
type III events, 14

unacceptable, 237
uncertain, 5, 273
uncertainty, 2, 7, 8, 10, 11, 22, 170, 190, 226, 229, 231, 242, 251, 253, 274
uncertainty management, 9
undesired, 43
undesired event, 123
Unknown Knowns, 14
unknowns, 252
Unknown Unknowns, 14
unsafe, 58
unsafe act, 175
unwanted event, 15, 148

variables, 141
Ventilation, 250
viability, 187
victims, 154
vulnerability, 124, 159, 167, 199
vulnerable, 18, 148

warning, 199
warning signs, 268
warning systems, 237
waste, 245
Weaknesses, 10, 98
witness, 179
workplace, 225
worsening factors, 18
worst-case, 136, 138

zero-accident, 77
Zero-risk, 17